Lecture Notes in Mathematics

Volume 2325

This series reports on new developments in all areas of mathematics and their applications - quickly, informally and at a high level. Mathematical texts analysing new developments in modelling and numerical simulation are welcome. The type of material considered for publication includes:

1. Research monographs
2. Lectures on a new field or presentations of a new angle in a classical field
3. Summer schools and intensive courses on topics of current research.

Texts which are out of print but still in demand may also be considered if they fall within these categories. The timeliness of a manuscript is sometimes more important than its form, which may be preliminary or tentative.

Titles from this series are indexed by Scopus, Web of Science, Mathematical Reviews, and zbMATH.

Dubravka Ban

p-adic Banach Space Representations

With Applications to Principal Series

Springer

Dubravka Ban
School of Mathematical and Statistical
Sciences
Southern Illinois University
Carbondale, IL, USA

ISSN 0075-8434   ISSN 1617-9692   (electronic)
Lecture Notes in Mathematics
ISBN 978-3-031-22683-0   ISBN 978-3-031-22684-7   (eBook)
https://doi.org/10.1007/978-3-031-22684-7

Mathematics Subject Classification: 22E50, 20G25, 11F70, 11F85, 46S10

This Springer imprint is published by the registered company Springer Nature Switzerland AG
The registered company address is: Gewerbestrasse 11, 6330 Cham, Switzerland

To the memory of my parents,
Anka and Nebomir

Preface

This book grew out of a course taught in Spring 2021 at Southern Illinois University. Its purpose is to lay the foundations of the representation theory of p-adic groups on p-adic Banach spaces, explain the duality theory of Schneider and Teitelbaum, and demonstrate its applications to continuous principal series. This monograph is intended to serve both as a reference book and as an introductory text for students entering the area. In addition, it could be of interest to mathematicians who are working in the representation theory on complex vector spaces and would like to learn more about p-adic Banach space representations.

The participants in the course were Devjani Basu, Jeremiah Roberts, Layla Sorkatti, Oneal Summers, An Tran, Manisha Varahagiri, and Menake Wijerathne. They prepared and presented lectures based on the first draft of the book. I would like to thank them for their patience in navigating through a half-finished book and for their corrections and comments.

Following the suggestions of the three referees, this monograph includes many improvements, broadening the scope of exposition. I would like to thank the referees for their detailed reviews and invaluable comments. Finally, I would like to thank Brian Conrad, Matthias Strauch, and Marie-France Vignéras for their contributions to the final version of the book.

Carterville, IL, USA   Dubravka Ban
September 2022

Contents

Chapter 1
Introduction

We study the representation theory of p-adic groups on p-adic Banach spaces. This is an active field of research whose foundations were laid by Peter Schneider and Jeremy Teitelbaum in [64].

We start with a sequence of finite field extensions $\mathbb{Q}_p \subseteq L \subseteq K$ and their rings of integers $\mathbb{Z}_p \subseteq o_L \subseteq o_K$. Our group G is the group of L-points of an algebraic group—a typical example is $G = GL_n(L)$. More specifically, $G = \mathbf{G}(L)$, where $\mathbf{G}$ is a split reductive $\mathbb{Z}$-group. We also consider $G_0 = \mathbf{G}(o_L)$, which is a maximal compact subgroup of G. We study continuous representations of G and G_0 on K-Banach spaces.

There are many types of representations of G. In general, a representation of G on a vector space V is a homomorphism $\pi : G \rightarrow \mathrm{Aut}(V)$. What we now call *classical representations of p-adic groups* are representations on complex vector spaces. The relevant category is the category of admissible-smooth representations, where *smooth* means locally constant and being *admissible-smooth* requires an additional finiteness condition (see Definition 6.12).

The topology on V does not play a role in the definition and properties of smooth representations. Hence, we can replace complex numbers by other fields. For instance, if ℓ is a prime number, we can consider ℓ-modular representations of G, which are smooth representations on vector spaces over the fields of characteristic ℓ. We just mention that the case $\ell = p$ differs significantly from $\ell \neq p$.

Enter K. When we look at the representations on K-vector spaces, it becomes immediately clear that there are many more interesting representations beyond smooth. For instance, we have algebraic representations, with the action of G described by polynomial functions. Algebraic representations of G are nice and natural, and most of them do not exist over $\mathbb{C}$ because there are no continuous polynomial functions $L \rightarrow \mathbb{C}$ except the constant ones. Similarly, we have locally-analytic representations of G, which again do not exist over $\mathbb{C}$.

The theory of representations of G on K-vector spaces branches into two lines of research: Banach space representations and locally analytic representations. The

© The Author(s), under exclusive license to Springer Nature Switzerland AG 2022
D. Ban, *p-adic Banach Space Representations*, Lecture Notes
in Mathematics 2325, https://doi.org/10.1007/978-3-031-22684-7_1

two lines are interconnected and interdependent, but the foundations are quite separated. In this book, we build the theory for Banach space representations and just occasionally mention locally analytic representations.

1.1 Admissible Banach Space Representations

When stepping into a new area, it is a challenge to decide what are the right objects to study. The category of K-Banach space representations carries certain pathologies (see Remark 4.36), and to avoid them, Schneider and Teitelbaum introduce in [64] an additional finiteness condition called admissibility. To explain it, we first introduce the Iwasawa algebra of G_0, defined as the projective limit

$$o_K[[G_0]] = \varprojlim_N o_K[G_0/N],$$

where N runs over the set of open normal subgroups of G_0. In addition, we define

$$K[[G_0]] = K \otimes_{o_K} o_K[[G_0]]$$

with the locally convex topology as in Definition 3.42. Then $K[[G_0]]$ can be identified with the convolution algebra $D^c(G_0, K)$ of continuous distributions on G_0 (Theorem 3.44).

Suppose that V is a K-Banach space representation of G, with continuous dual V'. Both V and V' carry the $K[[G_0]]$-module structures induced by the G-actions. The $K[[G_0]]$-module structure on V', together with the Schikhof duality [58], is a basis for the Schneider-Teitelbaum duality (Theorem 4.35).

The fundamental result by Lazard that $o_K[[G_0]]$ and $K[[G_0]]$ are noetherian rings [44] leads to the notion of admissibility. The K-Banach space representation V of G is called *admissible* if V' is finitely generated as a $K[[G_0]]$-module. We denote by

$$\text{Ban}_G^{\text{adm}}(K)$$

the category of all admissible K-Banach space representations of G. This category is algebraic in nature. Namely,

$$V \mapsto V' \tag{1.1}$$

defines an anti-equivalence between $\text{Ban}_{G_0}^{\text{adm}}(K)$ and the category of finitely generated $K[[G_0]]$-Iwasawa modules (Theorem 4.43). Finitely generated modules over a noetherian ring form a category with nice properties. By duality (1.1), these properties then also hold in $\text{Ban}_{G_0}^{\text{adm}}(K)$.

Behind the formulas and theorems, we recognize the beauty in the duality described above, justifying our choice to work with the category $\mathrm{Ban}_G^{\mathrm{adm}}(K)$. However, we have more to offer for justification. Most notably, admissible Banach space representations appear in Colmez' p-adic Langlands correspondence for $GL_2(\mathbb{Q}_p)$ [20, 22]. For us, it will be important that $\mathrm{Ban}_G^{\mathrm{adm}}(K)$ contains all continuous principal series. Thus, the category $\mathrm{Ban}_G^{\mathrm{adm}}(K)$ is rich in representations. The size and diversity of the category $\mathrm{Ban}_G^{\mathrm{adm}}(K)$ is discussed in Remark 4.51.

1.2 Principal Series Representations

Using the duality (1.1), we can study K-Banach space representations of G by considering the corresponding $K[[G_0]]$-modules. We apply this approach in Part II to study principal series representations.

Let $\mathbf{P}$ be a Borel subgroup of $\mathbf{G}$, having unipotent radical $\mathbf{U}$ and split maximal torus $\mathbf{T} \subset \mathbf{P}$. Denote by $\mathbf{U}^-$ the opposite subgroup of $\mathbf{U}$. Let $P = \mathbf{P}(L)$ and $P_0 = \mathbf{P}(o_L)$, and $U_0 = \mathbf{U}(o_L)$. If $\chi : P \to K^\times$ is a continuous character, we define

$$\mathrm{Ind}_P^G(\chi^{-1}) = \{f : G \to K \text{ continuous} \mid f(gp) = \chi(p)f(g) \text{ for all } p \in P, g \in G\}$$

with the action of G by left translations. We prove that $\mathrm{Ind}_P^G(\chi^{-1})$ has a natural structure as a Banach space representation of G (Proposition 7.5).

In this introduction, we use the same letter χ for $\chi|_{P_0} : P_0 \to o_K^\times$. Similarly as above, we define $\mathrm{Ind}_{P_0}^{G_0}(\chi^{-1})$. This is a Banach space with respect to the sup norm and the restriction map defines a topological isomorphism from $\mathrm{Ind}_P^G(\chi^{-1})$ to $\mathrm{Ind}_{P_0}^{G_0}(\chi^{-1})$ (Proposition 7.3).

The space $\mathrm{Ind}_P^G(\chi^{-1})$ can be described explicitly using a set of representatives of G/P. Let B be the standard Iwahori subgroup of G_0. We denote by $W = W(\mathbf{G}, \mathbf{T})$ the Weyl group of $\mathbf{G}$ relative to $\mathbf{T}$. For each $w \in W$ we select a representative $\dot{w} \in \mathbf{G}(\mathbb{Z})$. Let $U_{w,\frac{1}{2}}^- = \dot{w}^{-1}B\dot{w} \cap U_0^-$. Then we have the disjoint union decomposition

$$G = \coprod_{w \in W} \dot{w} U_{w,\frac{1}{2}}^- P$$

as in Proposition 5.45, which gives us the direct sum decomposition

$$\mathrm{Ind}_P^G(\chi^{-1}) \cong \bigoplus_{w \in W} C(U_{w,\frac{1}{2}}^-, K)$$

as in Proposition 7.3. Here, $C(U_{w,\frac{1}{2}}^-, K)$ is the Banach space of continuous functions $f : U_{w,\frac{1}{2}}^- \to K$ equipped with the sup norm.

The character χ extends to a character of $K[[P_0]]$. We denote by $K^{(\chi)}$ the $K[[P_0]]$-module structure on K induced by this character. We prove that the dual of $\mathrm{Ind}_P^G(\chi^{-1})$ is isomorphic to

$$M^{(\chi)} = K[[G_0]] \otimes_{K[[P_0]]} K^{(\chi)}$$

(see Theorem 7.12). As a $K[[G_0]]$-module, $M^{(\chi)}$ is generated by a single element $1 \otimes 1$, which implies that $\mathrm{Ind}_P^G(\chi^{-1})$ is admissible (Corollary 7.13).

Next, we want to use the duality (1.1) to obtain results about $\mathrm{Ind}_P^G(\chi^{-1})$. Our first step is to describe the structure of $M^{(\chi)}$. We give a projective limit realization of $M_0^{(\chi)} = o_K[[G_0]] \otimes_{o_K[[P_0]]} o_K^{(\chi)}$ (Proposition 7.20) and prove a $K[[B]]$-module decomposition

$$M^{(\chi_0)} \cong \bigoplus_{w \in W} K[[B]] \otimes_{K[[P_{\frac{1}{2}}^{w,\pm}]]} K^{(w\chi)}$$

where $P_{\frac{1}{2}}^{w,\pm} = B \cap w P_0 w^{-1}$ (Corollary 7.24).

In Chap. 8, we study intertwining operators on principal series representations. For any two continuous characters χ_1 and χ_2 of P, we want to compute the space

$$\mathcal{H}_{G_0}(\chi_1, \chi_2) = \mathrm{Hom}_{G_0}^c(\mathrm{Ind}_{P_0}^{G_0}(\chi_1^{-1}), \mathrm{Ind}_{P_0}^{G_0}(\chi_2^{-1}))$$

of continuous intertwining operators between $\mathrm{Ind}_{P_0}^{G_0}(\chi_1^{-1})$ and $\mathrm{Ind}_{P_0}^{G_0}(\chi_2^{-1})$. We first compute the space of $K[[G_0]]$-linear maps $\mathrm{Hom}_{K[[G_0]]}(M^{(\chi_1)}, M^{(\chi_2)})$ (see Corollary 8.13). Then we can use duality to get $\mathcal{H}_{G_0}(\chi_1, \chi_2)$. It turns out that it is equal to $\mathcal{H}_G(\chi_1, \chi_2)$, and we have

$$\mathcal{H}_G(\chi_1, \chi_2) = \mathcal{H}_{G_0}(\chi_1, \chi_2) = \begin{cases} 0 & \text{if } \chi_1 \neq \chi_2, \\ K \cdot \mathrm{id} & \text{if } \chi_1 = \chi_2 \end{cases} \tag{1.2}$$

(see Proposition 8.15). The description of $\mathcal{H}_G(\chi_1, \chi_2)$ in (1.2) may come as a surprise to a mathematician working with smooth representations of G. Still, the length of a smooth principal series of G is at most the order of the Weyl group, and the corresponding space of intertwining operators is always finite-dimensional. A full-blown surprise is the equality $\mathcal{H}_G(\chi_1, \chi_2) = \mathcal{H}_{G_0}(\chi_1, \chi_2)$ and the description of $\mathcal{H}_{G_0}(\chi_1, \chi_2)$. Here is why. Consider the case when $\chi_1 = \chi_2 = \chi$ is a smooth character. Let $V = \mathrm{Ind}_{P_0}^{G_0}(\chi^{-1})$ and let U be the smooth part of V. Then U is dense in V (see Lemma 7.7). As a G_0-representation, U decomposes as a countable direct sum of finite dimensional representations ρ with finite multiplicities $m(\rho)$:

$$U \cong \bigoplus_{\rho} m(\rho)\rho.$$

Each ρ is closed in V, so V contains countably many closed subrepresentations.

Clearly, U has numerous self-intertwining operators that are not scalar multiples of the identity. These operators, however, cannot be extended continuously to V. An example is worked out in Sect. 8.3.1.

1.3 Some Questions and Further Reading

Equality (1.2) illustrates how the theory of p-adic representations differs from the classical theory of smooth representations. It also shows that some of the standard methods from the classical theory are not suitable in this new context. For instance, intertwining operators are of no use for studying principal series on p-adic Banach spaces. Also, a serious difficulty comes from the lack of a p-adic Haar measure on G (see Sect. 3.4.2), thus making many of the integration-based methods inapplicable. Still, in both theories we can ask similar questions, despite the fact that the answers are often completely different.

The fundaments of the theory of smooth representations are well-established and can indicate interesting problems about p-adic Banach space representations. Cartier's survey [16] is a good place to start reading about admissible smooth representations, together with Casselman's lectures [17] and the first chapter of [14]. One of the concepts to investigate is parabolic induction. Namely, principal series representations are induced from a Borel subgroup, and we would like to know more about the induction from an arbitrary parabolic subgroup. For p-adic Banach space representations, this problem is mostly unexplored at the time of writing. Even for principal series, there is no comprehensive theory yet. Schneider's conjecture on irreducibility of continuous principal series and some related results are discussed in Sect. 8.4.

A true challenge will be to come up with replacements to integration-based methods. For instance, is there an analogue of the Harish-Chandra characters (distribution characters [36]) for p-adic Banach space representations?

An irreducible representation that does not appear as a subquotient of a parabolically induced representation is called supercuspidal. Construction and classification of smooth supercuspidal representations is a deep problem, addressed by the theory of types. We cannot but wonder what will be discovered about supercuspidal representations on p-adic Banach spaces. Answers, of course, are not just around the corner—what is hard in the smooth case can only get harder over Banach spaces.

We have a nice picture for unitary representations of $GL_2(\mathbb{Q}_p)$, where *unitary* means norm-preserving (see Sect. 4.4.1 for definitions). An admissible K-Banach space representation of $GL_2(\mathbb{Q}_p)$ is called *ordinary* if it is a subquotient of a continuous principal series induced from a unitary character. The classification of ordinary representation of $GL_2(\mathbb{Q}_p)$ is given in Sect. 8.2.1.

Non-ordinary representations of $GL_2(\mathbb{Q}_p)$ are covered by the p-adic Langlands correspondence, which is a correspondence between Galois representations and Banach space representations (see Remark 4.51 for a precise statement). The idea of the p-adic Langlands correspondence was introduced by Breuil in [11]. Berger's expository paper [6] is an excellent introduction to the p-adic Langlands correspondence for $GL_2(\mathbb{Q}_p)$. For other groups, see [15] and [12]. The correspondence for $GL_2(\mathbb{Q}_p)$ offers, among other things, an insight into the inner structure of the Banach space representations—particularly those containing smooth or locally algebraic vectors [7, 21]—thus indicating possible research directions for other groups.

Locally analytic vectors make an important part of the picture. Many of the results on p-adic Banach space representations rely on considerations of locally analytic vectors. An overview of some of the results and connections between the two theories can be found in [61]. For foundations of the theory of locally analytic representation, see [32] and [65]. Locally analytic principal series are well understood, thanks to the works of Orlik and Strauch [50, 51].

Finally, the theory of mod p representations can be useful for studying representations over K. Namely, if V is a unitary K-representation of G, then we can reduce it mod $\mathfrak{p}_K$ (which is briefly explained in Sect. 4.4.1) to obtain a smooth representation $\overline{V}$ over the residue field $\kappa = o_K/\mathfrak{p}_K$. As before, we consider two types of representations: components of parabolically induced representations and supercuspidal (also called supersingular) representations. The classification of smooth $\overline{\mathbb{F}}_p$-representations of G in terms of supercuspidals is given in [1]. As the authors put it, "By contrast, supercuspidal mod p representations remain a complete mystery, apart from the case of $GL_2(\mathbb{Q}_p)$ [11] and groups closely related to it." Still, as the theory of mod p representations develops, it will tell us more about unitary Banach space representations. How the two theories relate is beautifully illustrated by the compatibility of the p-adic Langlands correspondence for $GL_2(\mathbb{Q}_p)$ with reduction mod p [5].

1.4 Prerequisites

The only prerequisites for Part I are basic topology and some elementary properties of nonarchimedean fields (as summarized in Appendix A). The theory we present is built on projective limits and nonarchimedean functional analysis. Everything we need from these two areas is covered in the book. For projective limits, this is done at the beginning, in Sects. 2.1 and 2.2. For nonarchimedean functional analysis, we take a different approach (so that the introductory sections do not take too long). We cover it gradually, piece by piece, intertwined with the theory of Iwasawa algebras, continuous distributions, and Banach space representations. Sections 3.1, 3.3, and 4.2 are on general functional analysis, while the rest of Chaps. 3 and 4 deals with more specific topics directed towards explaining the admissible Banach space representations and the Schneider-Teitelbaum duality.

In Part II, we work with reductive $\mathbb{Z}$-groups. The structure theory of reductive groups is extensive and we give an overview (with no proofs) in Chap. 5. For the reader interested only in general linear groups, we summarize in Sect. 5.6 the structural components of GL_n needed in Chaps. 7 and 8.

1.5 Notation

In this book, p is a fixed prime number, and $\mathbb{Q}_p$ is the field of p-adic numbers. Throughout the book, K and L are finite extensions of $\mathbb{Q}_p$.

We denote by o_K the ring of integers of K and by $\mathfrak{p}_K$ its unique maximal ideal. We define o_L and $\mathfrak{p}_L$ similarly. The absolute value $|\ | = |\ |_K$ on K is given by $|\varpi_K| = q_K^{-1}$, where ϖ_K is a uniformizer of K and q_K is the cardinality of the residue field of K (see Appendix A.2.2).

If X is a set, 1_X denotes the characteristic function of X. We write $\mathbb{N}$ for the set of natural numbers, $\mathbb{N} = \{1, 2, 3, \dots\}$.

A more extensive notation is used in Part II, listed on page 89.

1.6 Groups

The groups will evolve throughout the book, because the theory we develop will impose more and more conditions.

G_0:   Throughout the book, G_0 is a compact group. In addition, it is

- profinite: in Chaps. 2, 3,
- a compact p-adic Lie group: in Chap. 4,
- the group of o_L-points of a split reductive $\mathbb{Z}$-group: in Part II.

G: The group G is

- a p-adic Lie group: in Chap. 4,
- the group of L-points of a split reductive $\mathbb{Z}$-group: in Part II.

Ultimately, we are interested in applying the theory on $G_0 = \mathbf{G}(o_L)$ and $G = \mathbf{G}(L)$, where $\mathbf{G}$ is a split connected reductive $\mathbb{Z}$-group. Then $\mathbf{G}(o_L)$ is a profinite group, and also a compact p-adic Lie group. Hence, any statement in the book that involves G_0 holds for $G_0 = \mathbf{G}(o_L)$. Even more specifically, it holds for $G_0 = GL_n(o_L)$.

Similarly, $\mathbf{G}(L)$ is a p-adic Lie group, and the results of Chap. 4 for G hold for $G = \mathbf{G}(L)$ and also for $G = GL_n(L)$.

Part I
Banach Space Representations of *p*-adic Lie Groups

Chapter 2
Iwasawa Algebras

Throughout the book, K and L are finite extensions of $\mathbb{Q}_p$.

In this chapter, G_0 is a profinite group. We define the Iwasawa algebra of G_0 (page 26) and study its properties. The Iwasawa algebra is defined using projective limits. Moreover, the group G_0 itself is defined as a projective limit. Hence, our first step in understanding Iwasawa algebras is to understand projective limits, first for topological spaces, and then for topological groups and o_K-modules.

2.1 Projective Limits

In this section, we define projective limits and describe some of their basic properties. Our presentation follows Chapter 1 in [54]. We are in particular interested in the projective limits of compact Hausdorff topological spaces—the kind of projective limit we will encounter in the definition of an Iwasawa algebra.

Definition 2.1 A **directed partially ordered set** or **directed poset** is a set I with a binary relation $\leq$ satisfying, for all $i, j, k \in I$,

  (i) $i \leq i$,
  (ii) if $i \leq j$ and $j \leq i$, then $i = j$, and
  (iii) if $i \leq j$ and $j \leq k$, then $i \leq k$.
  (iv) for any $i, j \in I$ there exists some $k \in I$ such that $i \leq k$ and $j \leq k$.

Definition 2.2 Let $(I, \leq)$ be a directed poset. An **inverse system** or **projective system** of topological spaces over I

$$(X_i, \varphi_{ij})_I$$

© The Author(s), under exclusive license to Springer Nature Switzerland AG 2022
D. Ban, *p-adic Banach Space Representations*, Lecture Notes
in Mathematics 2325, https://doi.org/10.1007/978-3-031-22684-7_2

is a family of topological spaces $\{X_i \mid i \in I\}$ together with continuous maps $\varphi_{ij} :$ $X_i \to X_j$, for all $i \geq j$, called the **connecting maps**, such that

(i) $\varphi_{ii} = \mathrm{id}_{X_i}$,
(ii) **Compatibility condition**: if $i \geq j \geq k$, then

$$\varphi_{jk} \circ \varphi_{ij} = \varphi_{ik},$$

that is, the following diagram commutes

$$X_i \xrightarrow{\ \varphi_{ij}\ } X_j \xrightarrow{\ \varphi_{jk}\ } X_k$$
$$\varphi_{ik}$$

We sometimes suppress φ_{ij} from notation and write simply $(X_i)_I$ or (X_i).

Example 2.3 Take the set of natural numbers $\mathbb{N}$ with standard $\leq$. For $n \in \mathbb{N}$, we consider the finite group $\mathbb{Z}/p^n\mathbb{Z}$ (with discrete topology). For $n \geq m$, let

$$\varphi_{n,m} : \mathbb{Z}/p^n\mathbb{Z} \to \mathbb{Z}/p^m\mathbb{Z}$$

be the reduction modulo p^m. The maps $\varphi_{n,m}$ clearly satisfy the compatibility condition, and hence $(\mathbb{Z}/p^n\mathbb{Z}, \varphi_{n,m})_{\mathbb{N}}$ is an inverse system.

Example 2.4 We equip $\mathbb{N}$ with the partial order $\preceq$ defined by

$$m \preceq n \quad \Longleftrightarrow \quad m \mid n.$$

For $n \in \mathbb{N}$, we consider the finite group $\mathbb{Z}/n\mathbb{Z}$ (with discrete topology). If $m \mid n$, let $\varphi_{n,m} : \mathbb{Z}/n\mathbb{Z} \to \mathbb{Z}/m\mathbb{Z}$ be the reduction modulo m. Similarly as in Example 2.3, $\varphi_{n,m}$ satisfy the compatibility condition, and $(\mathbb{Z}/n\mathbb{Z}, \varphi_{n,m})_{(\mathbb{N}, \preceq)}$ is an inverse system.

Definition 2.5 Let $(X_i, \varphi_{ij})_I$ be an inverse system of topological spaces and let Y be a topological space. Suppose that for every $i \in I$ we have a continuous map

$$\psi_i : Y \to X_i.$$

We say that the maps ψ_i are **compatible** if for every $i \geq j$

$$\varphi_{ij} \circ \psi_i = \psi_j,$$

that is, the following diagram commutes

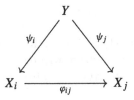

2.1.1 Universal Property of Projective Limits

Definition 2.6 Let $(X_i, \varphi_{ij})_I$ be an inverse system of topological spaces. A topological space X together with compatible continuous maps $\varphi_i : X \rightarrow X_i$ is said to be a **projective limit** or an **inverse limit** of $(X_i, \varphi_{ij})_I$ if the following **universal property** is satisfied:

- for any topological space Y equipped with compatible continuous maps $\psi_i : Y \rightarrow X_i$ there exists a unique continuous map $\psi : Y \rightarrow X$ such that

$$\varphi_i \circ \psi = \psi_i, \quad \text{for all } i \in I.$$

In this case, the following diagram commutes for all $i \geq j$

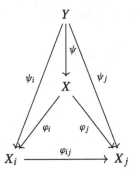

Theorem 2.7 *Let $(X_i, \varphi_{ij})_I$ be an inverse system of topological spaces. Then its projective limit exists and it is unique in the following sense: if $(X, \varphi_i)_I$ and $(X', \varphi_i')_I$ are two projective limits of $(X_i, \varphi_{ij})_I$, then there is a unique homeomorphism $\varphi : X \rightarrow X'$ such that $\varphi_i' \circ \varphi = \varphi_i$ for all $i \in I$.*

Proof For proving existence, we will construct a projective limit $(X, \varphi_i)_I$ as a subspace of the direct product $\prod_{i \in I} X_i$. Set

$$X = \{(x_i)_{i \in I} \in \prod_{i \in I} X_i \mid \varphi_{ij}(x_i) = x_j \text{ if } i \geq j\} \tag{2.1}$$

and for $j \in I$, define

$$\varphi_j((x_i)_{i \in I}) = x_j.$$

We equip $\prod_{i \in I} X_i$ with the product topology and X with the subspace topology. Then φ_j is continuous, because it is the restriction to X of the canonical projection $\prod_{i \in I} X_i \to X_j$. The requirement $\varphi_{ij}(x_i) = x_j$ from the definition of X assures that the maps φ_j are compatible.

Now, assume that Y is a topological space with compatible maps $\psi_i : Y \to X_i$. Define $\psi : Y \to \prod_{i \in I} X_i$ by

$$\psi(y) = (\psi_i(y))_{i \in I}.$$

Then ψ is continuous, by general properties of the product topology. The condition $\varphi_{ij} \circ \psi_i = \psi_j$ assures that the image of ψ is contained in X. Let us denote by the same letter ψ the corresponding map $\psi : Y \to X$ which is clearly continuous. The uniqueness of ψ follows from the condition $\varphi_i \circ \psi = \psi_i$.

The proof of uniqueness of a projective limit is standard, and it is left as an exercise. $\square$

If $(X_i, \varphi_{ij})_I$ is an inverse system, we denote its projective limit by

$$\varprojlim_{i \in I} X_i.$$

The maps $\varphi_i : \varprojlim_{i \in I} X_i \to X_i$ are called **projections** . They are not necessarily surjective. However, as we see from the proof of Theorem 2.7, they come from the canonical projections $\prod_{i \in I} X_i \to X_j$.

Example 2.8 Let $(\mathbb{Z}/p^n\mathbb{Z}, \varphi_{n,m})_\mathbb{N}$ be the inverse system defined in Example 2.3, where for $n \geq m$ the map $\varphi_{n,m} : \mathbb{Z}/p^n\mathbb{Z} \to \mathbb{Z}/p^m\mathbb{Z}$ is the reduction modulo p^m. Then

$$\varprojlim_{n \in \mathbb{N}} \mathbb{Z}/p^n\mathbb{Z} = \mathbb{Z}_p,$$

the ring of p-adic integers. We can identify $\mathbb{Z}/p^n\mathbb{Z}$ with $\mathbb{Z}_p/p^n\mathbb{Z}_p$, and the map

$$\varphi_n : \mathbb{Z}_p \to \mathbb{Z}/p^n\mathbb{Z} = \mathbb{Z}_p/p^n\mathbb{Z}_p,$$

for $n \in \mathbb{N}$, is the reduction modulo p^n.

Take $Y = \mathbb{Z}$ and let $\psi_n : \mathbb{Z} \to \mathbb{Z}/p^n\mathbb{Z}$ be the reduction modulo p^n. Then the diagram

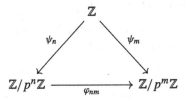

commutes for all $n \geq m$, so the maps ψ_n are compatible. By the universal property of projective limits, there exists the corresponding map $\mathbb{Z} \to \mathbb{Z}_p$. This map is actually the embedding $\iota : \mathbb{Z} \to \mathbb{Z}_p$.

2.1.2 Projective Limit Topology

Definition 2.9 Let X be a topological space. A family $\mathcal{B}$ of open subsets of X is called a **subbase of the topology** on X if all finite intersections of elements of $\mathcal{B}$ form a base of the topology on X.

Lemma 2.10 *Let $(X_i, \varphi_{ij})_I$ be an inverse system of topological spaces, and $X = \varprojlim_{i \in I} X_i$, with projection maps $\varphi_i : X \to X_i$. The sets*

$$\varphi_j^{-1}(U_j),$$

where $j \in I$ and U_j is an open subset of X_j, form a subbase of the topology on X.

Proof As in the proof of Theorem 2.7, we realize X as a topological subspace of the direct product $\prod_{i \in I} X_i$. For $j \in I$, denote by pr_j the canonical projection

$$\mathrm{pr}_j : \prod_{i \in I} X_i \to X_j.$$

If U_j is an open subset of X_j, then

$$\mathrm{pr}_j^{-1}(U_j) = \{(x_i)_{i \in I} \in \prod_{i \in I} X_i \mid x_j \in U_j\} \cong U_j \times \prod_{i \neq j} X_i$$

is an open set in $\prod_{i \in I} X_i$. The sets $\mathrm{pr}_j^{-1}(U_j)$ form a subbase of the topology on $\prod_{i \in I} X_i$. In this context, the projection map $\varphi_j : X \to X_j$ is just the restriction to X of pr_j. The sets

$$\varphi_j^{-1}(U_j) = \mathrm{pr}_j^{-1}(U_j) \cap X,$$

then form a subbase of the topology on X.   □

Lemma 2.11 *Let* $(X_i, \varphi_{ij})_I$ *be an inverse system of Hausdorff topological spaces, and* $X = \varprojlim_{i \in I} X_i$. *Then* X *is closed in* $\prod_{i \in I} X_i$.

Proof We will prove that the complement of X is open in $\prod_{i \in I} X_i$. Take $x = (x_i)_{i \in I}$ in $\prod_{i \in I} X_i$ and assume $x \notin X$. Then there exist $j, k \in I$, $j > k$, such that

$$\varphi_{jk}(x_j) \neq x_k.$$

In X_k, take the disjoint neighborhoods U_k of x_k and V_k of $\varphi_{jk}(x_j)$. By continuity of φ_{jk}, there exists an open neighborhood U_j of x_j in X_j such that $\varphi_{jk}(U_j) \subset V_k$. Define

$$U = \{(y_i)_{i \in I} \in \prod_{i \in I} X_i \mid y_j \in U_j, \ y_k \in U_k\} \cong U_j \times U_k \times \prod_{i \neq j,k} X_i.$$

This is an open neighborhood of x in $\prod_{i \in I} X_i$ disjoint from X.   □

Corollary 2.12 *If* $(X_i, \varphi_{ij})_I$ *is an inverse system of compact Hausdorff topological spaces, then* $\varprojlim_{i \in I} X_i$ *is also a compact Hausdorff topological space.*

Proof As a direct product of compact spaces, $\prod_{i \in I} X_i$ is compact. Then $\varprojlim_{i \in I} X_i$ is compact because it is a closed subspace of a compact space.   □

Proposition 2.13 *If* $(X_i, \varphi_{ij})_I$ *is an inverse system of compact Hausdorff nonempty topological spaces, then* $\varprojlim_{i \in I} X_i$ *is nonempty. In particular, the projective limit of an inverse system of nonempty finite sets is nonempty.*

Proof This is Proposition 1.1.4 in [54].   □

Cofinal Subsystem

Given an inverse system $(X_i, \varphi_{ij})_I$, we sometimes want to reduce the number of spaces X_i in a way that we still obtain the same projective limit.

Definition 2.14 Let $(I, \leq)$ be a directed poset and $J \subset I$ such that $(J, \leq)$ is also a directed poset. We say that J is **cofinal** in I if for every $i \in I$ there exists $j \in J$ such that $i \leq j$.

The following is Lemma 1.1.9 from [54].

Lemma 2.15 *Let* $(X_i, \varphi_{ij})_I$ *be an inverse system of compact topological spaces and let* J *be a cofinal subset of* I. *Then*

$$\varprojlim_{i \in I} X_i \cong \varprojlim_{j \in J} X_j.$$

Morphisms of Inverse Systems

Definition 2.16 Let $(X_i, \varphi_{ij})_I$ and $(Y_i, \psi_{ij})_I$ be two inverse systems of topological spaces indexed by the same directed poset I. A **morphism of inverse systems**

$$\theta : (X_i, \varphi_{ij})_I \to (Y_i, \psi_{ij})_I$$

is a family of continuous maps $\theta_i : X_i \to Y_i$ which are **compatible**, meaning that the following diagram commutes for all $i \geq j$

$$
\begin{array}{ccc}
X_i & \xrightarrow{\ \theta_i\ } & Y_i \\
{\scriptstyle \varphi_{ij}}\big\downarrow & & \big\downarrow{\scriptstyle \psi_{ij}} \\
X_j & \xrightarrow{\ \theta_j\ } & Y_j
\end{array}
$$

Suppose that we have a morphism of inverse systems $\theta = (\theta_i)_I : (X_i, \varphi_{ij})_I \to (Y_i, \psi_{ij})_I$. Let

$$X = \varprojlim_{i \in I} X_i \quad \text{and} \quad Y = \varprojlim_{i \in I} Y_i.$$

By the definition of the projective limit, we have the compatible maps $\varphi_i : X \to X_i$. Then $\theta_i \circ \varphi_i : X \to Y_i$ are compatible continuous maps. By the universal property of projective limits, we obtain the corresponding map $X \to Y$, denoted by $\varprojlim_{i \in I} \theta_i$. Hence,

$$\varprojlim_{i \in I} \theta_i : \quad \varprojlim_{i \in I} X_i \quad \to \quad \varprojlim_{i \in I} Y_i.$$

Lemma 2.17 *Let $\theta : (X_i, \varphi_{ij})_I \to (Y_i, \psi_{ij})_I$ be a morphism of inverse systems of topological spaces. If each component $\theta_i : X_i \to Y_i$ is injective, then*

$$\varprojlim_{i \in I} \theta_i : \quad \varprojlim_{i \in I} X_i \quad \to \quad \varprojlim_{i \in I} Y_i$$

is also injective.

Proof Let $X = \varprojlim_{i \in I} X_i$, $Y = \varprojlim_{i \in I} Y_i$, and $\theta = \varprojlim_{i \in I} \theta_i$. As in Eq. (2.1), we identify X with a subspace of $\prod_{i \in I} X_i$ and Y with a subspace of $\prod_{i \in I} Y_i$.

Suppose that x, x' are two different points in X. Write $x = (x_i)$ and $x' = (x_i')$. There exists $j \in I$ such that $x_j \neq x_j'$. Then $\theta_j(x_j) \neq \theta_j(x_j')$, because θ_j is injective. It follows $\theta(x) \neq \theta(x')$, thus proving injectivity of θ. $\qquad\square$

Example 2.18 The statement for surjective maps, analogous to Lemma 2.17, does not hold in general. Namely, let $\mathbb{Z}$ carry the usual discrete topology. Take the constant inverse system $(\mathbb{Z}, \mathrm{id})_{\mathbb{N}}$ and the inverse system $(\mathbb{Z}/p^n\mathbb{Z}, \varphi_{nm})_{\mathbb{N}}$ discussed in Example 2.8. Let

$$\theta_n : \mathbb{Z} \to \mathbb{Z}/p^n\mathbb{Z}$$

be the reduction modulo p^n. Then $(\theta_n)_{\mathbb{N}}$ is a morphism of inverse systems such that θ_n is surjective for all $n \in \mathbb{N}$. As discussed earlier,

$$\varprojlim_{n\in\mathbb{N}} \theta_i : \quad \mathbb{Z} = \varprojlim_{n\in\mathbb{N}} \mathbb{Z} \quad \to \quad \mathbb{Z}_p = \varprojlim_{n\in\mathbb{N}} \mathbb{Z}/p^n\mathbb{Z}$$

is the embedding $\iota : \mathbb{Z} \to \mathbb{Z}_p$. The group $\mathbb{Z}_p$ is compact because it is equal to the projective limit of compact groups (Corollary 2.12). On the other hand, $\mathbb{Z}$ is not compact. It follows that $\iota : \mathbb{Z} \to \mathbb{Z}_p$ is not surjective.

We will be working primarily with compact Hausdorff spaces, and for them, the following holds.

Lemma 2.19 *Let* $\Theta : (X_i, \varphi_{ij})_I \to (Y_i, \psi_{ij})_I$ *be a morphism of inverse systems of compact Hausdorff topological spaces. If each component* $\theta_i : X_i \to Y_i$ *is surjective, then*

$$\varprojlim_{i\in I} \theta_i : \quad \varprojlim_{i\in I} X_i \quad \to \quad \varprojlim_{i\in I} Y_i$$

is also surjective.

Proof Let $X = \varprojlim_{i\in I} X_i$, $Y = \varprojlim_{i\in I} Y_i$, and $\theta = \varprojlim_{i\in I} \theta_i$.

Take $y = (y_i) \in Y$. For each $i \in I$, let $\tilde{X}_i = \theta_i^{-1}(y_i)$. Observe that $\varphi_{ij}(\tilde{X}_i) \subseteq \tilde{X}_j$, so $(\tilde{X}_i)_I$ is an inverse system of topological spaces. Set

$$\tilde{X} = \varprojlim_{i\in I} \tilde{X}_i.$$

Then $\tilde{X} \subset X$ and if $\tilde{x} \in \tilde{X}$, then $\theta(\tilde{x}) = y$, by construction of $\tilde{X}$. Hence, to prove surjectivity, it suffices to show that $\tilde{X}$ is nonempty. For this, we first observe that each $\tilde{X}_i$ is compact, as a closed subspace of the compact set X_i. Then Proposition 2.13 implies that $\tilde{X} \neq \emptyset$. □

Corollary 2.20 *Let* $(X_i, \varphi_{ij})_I$ *be an inverse system of compact Hausdorff spaces and let* Y *be a compact Hausdorff space. Suppose that* $\theta_i : Y \to X_i$ *are compatible continuous surjective maps. Then the corresponding map* $\theta : Y \to \varprojlim_{i\in I} X_i$ *is surjective.*

Proof Apply Lemma 2.19 on the constant inverse system $(Y, \mathrm{id}_Y)_I$. □

Proposition 2.21 *Let $(X_i, \varphi_{ij})_I$ be an inverse system of compact Hausdorff spaces. Let $X = \varprojlim_{i \in I} X_i$, with the projections $\varphi_i : X \to X_i$. If Y is a subspace of X, then*

(i) *Y is dense in $\varprojlim_{i \in I} \varphi_i(Y)$.*
(ii) *If Y is a closed subspace of X, then $Y = \varprojlim_{i \in I} \varphi_i(Y)$.*
(iii) *Suppose that $\varphi_i(Y)$ is closed in X_i, for all i. Then*

$$\overline{Y} = \varprojlim_{i \in I} \varphi_i(Y),$$

  where $\overline{Y}$ is the closure of Y in X.

Proof The statements for empty spaces clearly hold, so we will assume that both X and Y are nonempty. Set $Y_i = \varphi_i(Y)$. Then $Y \subseteq \varprojlim_{i \in I} Y_i$.

(i) Take $y_0 \in \varprojlim_{i \in I} Y_i$. As in the proof of Theorem 2.7, we realize X as a subspace of $\prod_{i \in I} X_i$. Then $y_0 = (y_i)_{i \in I} \in \prod_{i \in I} Y_i$. Take an arbitrary neigborhood of y_0 in $\varprojlim_{i \in I} Y_i$. It contains an open neighborhood of y_0 of the form

$$U = \left(\prod_{j \in J} U_j \times \prod_{i \in I \setminus J} Y_i \right) \cap \varprojlim_{i \in I} Y_i$$

where J is a finite subset of I and U_j is an open neighborhood of y_j in Y_j. Select an index $k \in I$ such that $j \leq k$ for all $j \in J$ and choose $y \in Y$ such that $\varphi_k(y) = y_k$. Then $y \in U$, thus proving density. Assertion (ii) follows immediately from (i).

(iii) If Y_i is closed in X_i, for all i, then $\prod_{i \in I} Y_i$ is closed in $\prod_{i \in I} X_i$. It follows that $\varprojlim_{i \in I} Y_i$ is closed in X. By density (i), it follows that $\overline{Y} = \varprojlim_{i \in I} \varphi_i(Y)$.   □

2.2   Projective Limits of Topological Groups and o_K-Modules

An inverse system of topological spaces can carry additional algebraic structures. Recall that a **topological group** is a group G together with a topology on G such that both the product $G \times G \to G$ and the inverse map $G \to G, g \mapsto g^{-1}$, are continuous. Here, $G \times G$ carries the product topology. Similarly, a **topological ring** is a ring R together with a topology on R such that both the addition and multiplication are continuous maps $R \times R \to R$, where $R \times R$ carries the product topology. We will work with the following two:

• An **inverse system of topological groups** is an inverse system of topological spaces $(H_i, \varphi_{ij})_I$ such that each H_i is a topological group and that the connecting maps φ_{ij} are group homomorphisms. The projective limit

$$H = \varprojlim_{i \in I} H_i$$

has a natural group structure and, with the projective limit topology, it is a topological group (Exercise 2.22). The projection maps $\varphi_i : H \to H_i$ are group homomorphisms, and they are called **projection homomorphisms** .

- We define similarly an **inverse system of topological rings** and **inverse system of topological o_K-modules** .

Exercise 2.22 Let $(H_i, \varphi_{ij})_I$ be an inverse system of topological spaces such that each H_i is a topological group and that the connecting maps φ_{ij} are group homomorphisms. Let $H = \varprojlim_{i \in I} H_i$ be the topological projective limit. Prove that there is a unique group structure on H such that the projection maps $\varphi_i : H \to H_i$ are group homomorphisms for all $i \in I$.

Exercise 2.23 Formulate the statement for topological rings and topological o_K-modules as in Exercise 2.22. Prove it.

A sequence

$$0 \to (A_i)_I \to (B_i)_I \to (C_i)_I \to 0$$

of inverse systems of topological groups (respectively, topological o_K-modules) is said to be **exact** if the corresponding sequence of group homomorphisms (respectively, homomorphisms of topological o_K-modules)

$$0 \to A_i \to B_i \to C_i \to 0$$

is exact for all $i \in I$.

Proposition 2.24 *Let* $0 \to (A_i)_I \to (B_i)_I \to (C_i)_I \to 0$ *be an exact sequence of inverse systems of topological groups (respectively, topological o_K-modules). Then*

$$0 \to \varprojlim_{i \in I} A_i \to \varprojlim_{i \in I} B_i \to \varprojlim_{i \in I} C_i$$

is exact.

Proof The proof is left as an exercise. Injectivity follows from Lemma 2.17. □

Remark 2.25 We can consider the projective limit as a functor between appropriate categories. Proposition 2.24 tells us that $\varprojlim_{i \in I}$ is left exact on compact topological groups and compact topological o_K-modules. There is a condition, called the Mittag-Leffler condition, which assures that the functor $\varprojlim_{i \in I}$ is exact (see [43, Proposition 10.3]). For us, it will be important that exactness holds for compact Hausdorff topological groups and compact Hausdorff topological o_K-modules (see Proposition 2.26 below).

Proposition 2.26 *Let* $0 \to (A_i)_I \to (B_i)_I \to (C_i)_I \to 0$ *be an exact sequence of inverse systems of compact Hausdorff topological groups (respectively, compact Hausdorff topological o_K-modules). Then*

$$0 \to \varprojlim_{i \in I} A_i \to \varprojlim_{i \in I} B_i \to \varprojlim_{i \in I} C_i \to 0$$

is exact.

Proof Follows from Proposition 2.24 and Lemma 2.19. □

We have already learned that the inverse systems built on the projections $\mathbb{Z} \to \mathbb{Z}/p^n\mathbb{Z}$ show all kind of things that can go wrong in the projective limit, among them the exact sequence in the exercise below.

Exercise 2.27 Prove that the exact sequences $0 \to p^n\mathbb{Z} \to \mathbb{Z} \to \mathbb{Z}/p^n\mathbb{Z} \to 0$, for $n \in \mathbb{N}$, are compatible with respect to canonical maps. Hence, they define the exact sequence of inverse systems

$$0 \to (p^n\mathbb{Z})_\mathbb{N} \to (\mathbb{Z})_\mathbb{N} \to (\mathbb{Z}/p^n\mathbb{Z})_\mathbb{N} \to 0, \tag{2.2}$$

where $(\mathbb{Z})_\mathbb{N}$ is the constant inverse system $(\mathbb{Z}, \mathrm{id}_\mathbb{Z})_\mathbb{N}$. Show that the corresponding sequence of projective limits is

$$0 \to 0 \to \mathbb{Z} \to \mathbb{Z}_p.$$

As discussed in Example 2.18, the embedding $\mathbb{Z} \to \mathbb{Z}_p$ is not surjective. Hence, applying the projective limit on the short exact sequence (2.2) does not result in a short exact sequence. It follows that the projective limit functor is not right exact.

2.2.1 Profinite Groups

Definition 2.28

(i) A **profinite group** is a topological group that is isomorphic to the projective limit of an inverse system of discrete finite groups.

(ii) A **pro-p group** is a topological group that is isomorphic to the projective limit of an inverse system of discrete finite p-groups.

Example 2.29

(a) Consider the p-adic integers $\mathbb{Z}_p$ as an abelian group. Since

$$\mathbb{Z}_p = \varprojlim_{n \in \mathbb{N}} \mathbb{Z}/p^n\mathbb{Z},$$

$\mathbb{Z}_p$ is a profinite and pro-p group.

(b) Similarly, let o_L be the ring of integers in the p-adic field L and let $\mathfrak{p}_L$ be the unique maximal ideal in o_L. Then [49, Proposition 4.5]

$$o_L = \varprojlim_{n \in \mathbb{N}} o_L / \mathfrak{p}_L^n.$$

It follows that o_L is a pro-p group (see Exercise 2.31).

(c) Let us denote by $o_L^\times$ the group of units in o_L. For $n \in \mathbb{N}$, define $U^{(n)} = 1 + \mathfrak{p}_L^n$. This is a subgroup of the multiplicative group $o_L^\times$. Then [49, Proposition 4.5]

$$o_L^\times = \varprojlim_{n \in \mathbb{N}} o_L^\times / U^{(n)}.$$

The group $o_L^\times$ is profinite, but not pro-p (see Exercise 2.31).

Exercise 2.30 Using literature in algebraic number theory, find the orders of the following groups:
  (a) $o_L / \mathfrak{p}_L^n$,   (b) $o_L^\times / U^{(n)}$,   (c) $U^{(n)} / U^{(n+1)}$.

Exercise 2.31

(a) Let G be a profinite group. Suppose that G contains an element g of finite order such that p does not divide $|g|$. Prove that G is not a pro-p group.
(b) Prove that $o_L^\times$ is not a pro-p group. (This could be done using (a).)

Topology on Profinite Groups

For our applications, it is convenient to denote the profinite group in the proposition below by G_0, and the groups introduced in assertion (ii) by G_i (see Examples 2.35 and 2.37). To avoid confusion, we assume either $0 \notin I$ or $H_0 = 1$.

Proposition 2.32 *Let $(H_i, \varphi_{ij})_I$ be an inverse system of discrete finite groups and let*

$$G_0 = \varprojlim_{i \in I} H_i,$$

with projection homomorphisms $\varphi_i : G_0 \to H_i$. Then

  (i) *G_0 is a compact Hausdorff group.*
 (ii) *For any $i \in I$, the group*

$$G_i = \ker \varphi_i$$

is a normal compact open subgroup of G_0.
(iii) *The groups G_i, $i \in I$, form a fundamental system of open neighborhoods of the identity in G_0.*

(iv) Let $N(G_0)$ be the set of all open normal subgroups of G_0. Then

$$G_0 \cong \varprojlim_{N \in N(G_0)} G_0/N \cong \varprojlim_{i \in I} G_0/G_i.$$

Proof

(i) Follows from Corollary 2.12, because H_i are compact Hausdorff groups.

(ii) As the kernel of a continuous homomorphism into a Hausdorff group, G_i is a closed normal subgroup of G_0. Compactness of G_0 implies that G_i is compact as well. It is open because it is precisely an open subset of G of the form

$$G_i = \varphi_i^{-1}(1)$$

as described in Lemma 2.10. Here, 1 is the trivial subgroup of H_i, and it is an open subgroup in the discrete group H_i.

(iii) It follows from Lemma 2.10 that G_0 has a fundamental system of open neighborhoods of the identity consisting of the subgroups of the form $G_{i_1} \cap \cdots \cap G_{i_s}$ where $\{i_1, \ldots, i_s\}$ is a finite subset of I. Take $j \in I$ such that $j \geq i_k$ for all $i_k \in \{i_1, \ldots, i_s\}$. Then G_j is an open subgroup of $G_{i_1} \cap \cdots \cap G_{i_s}$.

(iv) The projections $G_0 \to G_0/G_i$ give us, using the universal property of projective limits, a continuous homomorphism $\theta : G_0 \to \varprojlim_{i \in I} G_0/G_i$. Corollary 2.20 tells us that θ is surjective. To find its kernel, we realize G_0 as a subset of $\prod_i H_i$, as in Eq. (2.1). Then

$$\ker \theta = \{(h_i)_{i \in I} \in G_0 \mid (h_i)_{i \in I} \in G_j \, \forall j \in I\} = \{(h_i)_{i \in I} \in G_0 \mid h_i = 1 \, \forall i \in I\} = 1.$$

It follows $G_0 \cong \varprojlim_{i \in I} G_0/G_i$. To show that $\varprojlim_{i \in I} G_0/G_i \cong \varprojlim_{N \in N(G_0)} G_0/N$, we observe that $(G_i)_I$ is cofinal in $N(G_0)$, and apply Lemma 2.15. $\square$

Exercise 2.33 Prove that the projective limit topology on o_L coincides with the topology induced by the nonarchimedean absolute value on L. Prove that the ideals $\mathfrak{p}_L^n$ form a neighborhood basis of 0 in o_L.

Exercise 2.34 Prove that the projective limit topology on $o_L^\times$ coincides with the topology induced by the nonarchimedean absolute value on L. Prove that the subgroups $U^{(n)} = 1 + \mathfrak{p}_L^n$ form a neighborhood basis of 1 in $o_L^\times$.

Example 2.35 Let $G = GL_n(L)$ be the group of $n \times n$ matrices with coefficients in L. Let $G_0 = GL_n(o_L)$ be the subgroup of $GL_n(L)$ consisting of all matrices $g = (g_{ij})$ such that the coefficients $g_{ij} \in o_L$ and $\det g \in o_L^\times$. For $n \in \mathbb{N}$, define

$$G_n = \{g \in G_0 \mid g \equiv 1 \mod \mathfrak{p}_L^n\}.$$

Then G_n is a normal subgroup of G_0, G_0/G_n is finite, and

$$G_0 = \varprojlim_{n \in \mathbb{N}} G_0/G_n.$$

We consider G_0 equipped with the projective limit topology (which is by Exercise 2.36 equal to the standard topology on G_0). The groups G_n, $n \in \mathbb{N}$, are compact and open, and they form a neighborhood basis of 1 in G.

Exercise 2.36 Let $M_{n \times n}(L)$ be the space of $n \times n$ matrices with coefficients in L. Then $M_{n \times n}(L)$ is an n^2-dimensional L-vector space and it is equipped with the standard norm: if $g = (g_{ij}) \in M_{n \times n}(L)$, then

$$\|g\| = \max_{i,j} |g_{ij}|.$$

Prove that the topology on G_0 induced by this norm is equal to the projective limit topology.

The following example is central for the theory developed in Chaps. 7 and 8.

Example 2.37 Similarly to Example 2.35 for GL_n, we can consider a general split reductive $\mathbb{Z}$-group $\mathbf{G}$. Such groups will be explained in detail in Sect. 5.5. Here, we just mention their connection with projective limits. Let $G_0 = \mathbf{G}(o_L)$ and let G_n be the kernel of $G_0 \to \mathbf{G}(o_L/\mathfrak{p}_L^n)$. Then G_n is a normal subgroup of G_0, G_0/G_n is finite, and

$$G_0 = \varprojlim_{n \in \mathbb{N}} G_0/G_n.$$

We equip G_0 with the projective limit topology. The groups G_n, $n \in \mathbb{N}$, are compact and open, and they form a neighborhood basis of 1 in G (see Lemma 5.40).

Exercise 2.38 Let G_0 be a profinite group, and let N be an open subgroup of G_0. Prove that G_0/N is finite.

Exercise 2.39 A nonempty topological space X is said to be **totally disconnected** if the connected components in X are the one-point sets. Let G_0 be a profinite group. Prove that G_0 is totally disconnected.

2.3   Iwasawa Rings

We are now ready to define Iwasawa rings. An important property of Iwasawa rings is that their topology is algebraic in nature, given by ideals. This falls under the umbrella of linear-topological modules defined below.

2.3.1 Linear-Topological o_K-Modules

Definition 2.40 A topological o_K-module is called **linear-topological** if the zero element has a fundamental system of open neighborhoods consisting of o_K-submodules.

Example 2.41 The ring o_K is a linear-topological o_K-module because the ideals $\mathfrak{p}_K^n$ form a fundamental system of open neighborhoods of zero.

Lemma 2.42 *Let $H = \{h_1, \ldots, h_n\}$ be a finite group. The group ring*

$$o_K[H] = \{a_1h_1 + \cdots + a_nh_n \mid a_i \in o_K\}$$

has the natural topology as a free o_K-module of rank n. With this topology, $o_K[H]$ is a topological ring and a linear-topological o_K-module. The ideals $\mathfrak{p}_K^m[H]$, $m \in \mathbb{N}$, form a neighborhood basis of zero.

Proof As an o_K-module, $o_K[H] \cong o_K^n$, and the topology on $o_K[H]$ is defined as the product topology on o_K^n. It is then easy to show that $o_K[H]$ is a topological ring and a topological o_K-modules.

To show that the ideals $\mathfrak{p}_K^m[H]$, $m \in \mathbb{N}$, form a neighborhood basis of zero, we first observe that for every m, $\mathfrak{p}_K^m$ is open in o_K. Then $\mathfrak{p}_K^m \times \cdots \times \mathfrak{p}_K^m$ is open in o_K^n and hence $\mathfrak{p}_K^m[H]$ is open in $o_K[H]$.

Let U be an open neighborhood of zero in $o_K[H]$. Then there exist open neighborhoods of zero $U_1, \ldots, U_n \subset o_K$ such that $U_1h_1 + \cdots + U_nh_n \subset U$. For each i, there exists m_i such that $\mathfrak{p}_K^{m_i} \subset U_i$. Let $m = \max_i\{m_i\}$. Then $\mathfrak{p}_K^m \subset U_i$, for all $i = 1, \ldots, n$, and

$$\mathfrak{p}_K^m[H] = \mathfrak{p}_K^m h_1 + \cdots + \mathfrak{p}_K^m h_n \subset U_1h_1 + \cdots + U_nh_n \subset U.$$

This proves that the ideals $\mathfrak{p}_K^m[H]$ form a neighborhood basis of zero in $o_K[H]$, thus also proving that $o_K[H]$ is a linear-topological o_K-module. $\square$

Definition of Iwasawa Algebra

Let G_0 be a profinite group. From Proposition 2.32, it follows that we can write G_0 as

$$G_0 \cong \varprojlim_{n \in \mathcal{N}} G_0/G_n,$$

where $\mathcal{N}$ is a directed poset and $\{G_n \mid n \in \mathcal{N}\}$ is a family of compact open subgroups of G_0 such that $G_n \leq G_m$ for $n \geq m$. For instance, we may take $\mathcal{N}$ to be the indexing set for all open normal subgroups of G_0. It is convenient to assume $G_0 \in \{G_n \mid n \in \mathcal{N}\}$, with the index $0 \in \mathcal{N}$.

Fix $n \in \mathcal{N}$. Then G_0/G_n is finite by Exercise 2.38, and the group ring $o_K[G_0/G_n]$ is a topological ring and a linear-topological o_K-module, as described in Lemma 2.42.

If $n, m \in \mathcal{N}$ and $n \geq m$, we denote by $\varphi_{n,m}$ the natural projection $\varphi_{n,m}$: $o_K[G_0/G_n] \to o_K[G_0/G_m]$. More specifically, if $\{g_1, \ldots, g_t\}$ is a set of coset representatives of G_0/G_n, an element of $o_K[G_0/G_n]$ can be written as $a_1 g_1 G_n + \cdots + a_t g_t G_n$. Then

$$\varphi_{n,m}(a_1 g_1 G_n + \cdots + a_t g_t G_n) = a_1 g_1 G_m + \cdots + a_t g_t G_m.$$

The maps $\varphi_{n,m}$ are clearly compatible. Hence,

$$(o_K[G_0/G_n], \varphi_{n,m})_{\mathcal{N}}$$

is an inverse system of topological rings and o_K-modules. Define

$$o_K[[G_0]] = \varprojlim_{n \in \mathcal{N}} o_K[G_0/G_n].$$

We equip $o_K[[G_0]]$ with the projective limit topology. Then $o_K[[G_0]]$ is a topological ring and a topological o_K-module. Clearly, it is torsion-free as an o_K-module. As a projective limit of compact rings, $o_K[[G_0]]$ is compact. The ring $o_K[[G_0]]$ is called the **completed group ring** or the **Iwasawa algebra** of G_0 over o_K. It was introduced and studied by Lazard in [44].

As in Eq. (2.1), we can realize $o_K[[G_0]]$ as a subset of $\prod_{n \in \mathcal{N}} o_K[G_0/G_n]$. Then we can write $\mu \in o_K[[G_0]]$ as

$$\mu = (\mu_n)_{n \in \mathcal{N}}, \quad \mu_n \in o_K[G_0/G_n].$$

Let $m, n \in \mathcal{N}$ such that $n > m$. We select a set of coset representatives $\{g_1, \ldots, g_s\}$ of G_0/G_m, and a set of coset representatives $\{h_1, \ldots, h_r\}$ of G_m/G_n. Then

$$\{g_i h_j \mid 1 \leq i \leq s, \ 1 \leq j \leq r\}$$

is a set of coset representatives of G_0/G_n. We can write

$$\mu_m = \sum_{i=1}^{s} a_i g_i G_m \quad \text{and} \quad \mu_n = \sum_{i=1}^{s} \sum_{j=1}^{r} b_{ij} g_i h_j G_n.$$

Then $\varphi_{n,m}(\sum_{j=1}^{r} b_{ij} g_i h_j G_n) = a_i g_i G_m$, and hence

$$\sum_{j=1}^{r} b_{ij} = a_i, \quad \text{for all } i \in \{1, \ldots, s\}.$$

The map $\varphi_{n,m}$ can be represented by the following diagram:

$$\mu_n = \cdots + \underbrace{b_{i1}g_i h_1 G_n + b_{i2}g_i h_2 G_n + \cdots b_{ir}g_i h_r G_n} + \cdots$$

$$\downarrow$$

$$\mu_m = \cdots + \qquad \left(\sum_{j=1}^{r} b_{ij}\right) g_i G_m \qquad\qquad + \cdots$$

Fundamental System of Neighborhoods of Zero

Lemma 2.43 *For $n \in N$, let φ_n be the projection homomorphism $o_K[[G_0]] \to o_K[G_0/G_n]$, and for $m \in \mathbb{N}$, let pr_m denote the projection $\mathrm{pr}_m : o_K \to o_K/\mathfrak{p}_K^m$ and also the induced map $\mathrm{pr}_m : o_K[G_0/G_n] \to o_K/\mathfrak{p}_K^m[G_0/G_n]$.*

(i) The two-sided ideals

$$J_{m,n}(G_0) = \ker(\mathrm{pr}_m \circ \varphi_n),$$

for $m \in \mathbb{N}$ and $n \in N$, form a fundamental system of open neighborhoods of zero in $o_K[[G_0]]$. In particular, $o_K[[G_0]]$ is a linear-topological o_K-module.

(ii) The ideal

$$\mathfrak{m}(G_0) = J_{1,0}(G_0)$$

is a maximal ideal in $o_K[[G_0]]$.

Proof From Lemma 2.10, we know that the sets

$$\varphi_n^{-1}(U_n),$$

where $n \in N$ and U_n is an open set in $o_K[G_0/G_n]$, form a subbase of the topology on $o_K[[G_0]]$. By Lemma 2.42, any neighborhood of zero in $o_K[G_0/G_n]$ contains a neighborhood of the form $\mathfrak{p}_K^m[G_0/G_n]$. Since

$$\varphi_n^{-1}(\mathfrak{p}_K^m[G_0/G_n]) = J_{m,n}(G_0),$$

it follows that finite intersections of the ideals $J_{m,n}(G_0)$, for $m \in \mathbb{N}$ and $n \in N$, form a fundamental system of open neighborhoods of zero in $o_K[[G_0]]$.

Now, let $J_1 = J_{m_1,n_1}(G_0)$ and $J_2 = J_{m_2,n_2}(G_0)$. Set $n \geq \{n_1, n_2\}$ and $m = \max\{m_1, m_2\}$. Then

$$J_{m,n}(G_0) \subset J_1 \cap J_2.$$

This proves that the ideals $J_{m,n}(G_0)$, for $m \in \mathbb{N}$ and $n \in \mathcal{N}$, actually form a fundamental system of open neighborhoods of zero in $o_K[[G_0]]$.

(ii) Since $o_K[[G_0]]/\mathfrak{m}(G_0) \cong o_K/\mathfrak{p}_K$, the ideal $\mathfrak{m}(G_0)$ is maximal.   □

Embedding $o_K[G_0]$, G_0, and o_K into $o_K[[G_0]]$

Lemma 2.44 *The canonical projections* $\theta_n : o_K[G_0] \to o_K[G_0/G_n]$ *are compatible and induce in the limit an injective ring homomorphism*

$$\theta : o_K[G_0] \to o_K[[G_0]]$$

The image of θ *is dense in* $o_K[[G_0]]$.

We use this homomorphism to identify $o_K[G_0]$ with its image in $o_K[[G_0]]$, endowed with the subspace topology of $o_K[[G_0]]$.

Proof The projections θ_n are clearly compatible, so by the universal property of projective limits, there exists a continuous ring homomorphism $\theta : o_K[G_0] \to o_K[[G_0]]$. To show that θ is injective, take a nonzero element $\mu \in o_K[G_0]$,

$$\mu = a_1 g_1 + \cdots + a_s g_s,$$

where $a_i \in o_K \setminus \{0\}$ and $g_i \in G$. There exists $n \in \mathcal{N}$ such that $g_i G_n \cap g_j G_n = \emptyset$, for all $i \neq j$. Then $\theta_n(\mu) \neq 0$, and $\theta(\mu) \neq 0$.

Finally, Proposition 2.21 (iii) tells us that the image of θ is dense in $o_K[[G_0]]$, because the projections $\theta_n : o_K[G_0] \to o_K[G_0/G_n]$ are surjective.   □

Since $o_K[G_0]$ is dense in $o_K[[G_0]]$, for every $\mu \in o_K[[G_0]]$ and every $J_{m,n}(G_0)$ there exists $\eta \in o_K[G_0]$ such that $\mu - \eta \in J_{m,n}(G_0)$. We can construct such η explicitly. Write

$$\mu = (\mu_n)_{n=1}^{\infty} \in \prod_{n \in \mathcal{N}} o_K[G_0/G_n].$$

We select a set of representatives $\{g_1, \ldots, g_s\}$ of G_0/G_n. Then we can write

$$\mu_n = a_1 g_1 G_n + \cdots + a_s g_s G_n,$$

where $a_1, \ldots, a_s \in o_K$. Define

$$\eta = a_1 g_1 + \cdots + a_s g_s.$$

Then $\mu - \eta \in J_{m,n}(G_0)$. Moreover, $\mu - \eta \in J_{\ell,n}(G_0)$, for all $\ell \in \mathbb{N}$. Hence, we have proved.

Lemma 2.45 *Let* $n \in \mathcal{N}$ *and let* $\mathcal{J}_n$ *be the kernel of the canonical projection* $o_K[G_0] \to o_K[G_0/G_n]$. *Then*

(i) $\mathcal{J}_n = \bigcap_{m \in \mathbb{N}} J_{m,n}$.
(ii) For every $\mu \in o_K[[G_0]]$ *there exists* $v \in o_K[G_0]$ *such that* $\mu - v \in \mathcal{J}_n$.

Exercise 2.46 Given $\mu \in o_K[[G_0]]$, construct explicitly a sequence $\eta^{(n)}$ in $o_K[G_0]$ such that $\lim_{n\to\infty} \eta^{(n)} = \mu$. (*Let* X *be a topological space, and let* x_n, $n \in \mathbb{N}$ *be a sequence in* X. *We say that* x_n *converges to* $x \in X$ *if for every neighborhood* U *of* x *there exists* $n_U \in \mathbb{N}$ *such that* $x_n \in U$ *for every* $n \geq n_U$.)

Since $G_0 \subset o_K[G_0]$, the embedding $o_K[G_0] \hookrightarrow o_K[[G_0]]$ from Lemma 2.44 gives us the inclusion $G_0 \hookrightarrow o_K[[G_0]]$.

Lemma 2.47 *The inclusion* $G_0 \hookrightarrow o_K[[G_0]]$ *is a homeomorphism onto its image.*

Proof The set $\{1 + J_{m,n}(G_0) \mid m \in \mathbb{N}, n \in \mathcal{N}\}$ is a neighborhood basis of 1 in $o_K[[G_0]]$. Since $m \geq 1$, we have

$$G_0 \cap (1 + J_{m,n}(G_0)) = G_n.$$

The lemma then follows immediately, because $\{G_n \mid m \in \mathbb{N}\}$ is a neighborhood basis of 1 in G_0. $\qquad\square$

It is interesting to notice that the intersection $G_0 \cap (1 + J_{m,n}(G_0)) = G_n$ is same for all $m \geq 1$.

It is customary to denote by 1 both the identity in o_K and the identity in G_0, and we will mostly do the same. At few places, however, we will denote the identity in G_0 by e.

We have the standard embedding $o_K \hookrightarrow o_K[G_0]$ given by $a \mapsto a \cdot 1$. We write simply a for $a \cdot 1$. In particular, $1 = 1 \cdot 1$ (or $1 = 1 \cdot e$), where

$$\underbrace{1}_{\in o_K[[G_0]]} = \underbrace{1}_{\in o_K} \cdot \underbrace{1}_{\in G_0}$$

Also, for $g \in G_0$, we write $-g$ for $(-1)g$. This notation makes sense because $(-1)g$ is the additive inverse of g in the ring $o_K[[G_0]]$. However, this should not be confused with matrix operations. Suppose that G_0 is a group of matrices, such as $GL_n(o_K)$. For a matrix $g = (a_{ij}) \in G_0$, the element $-g = (-1)g \in o_K[[G_0]]$ is not equal to the matrix $(-a_{ij})$. Similarly, for $c \in o_K$, the element $cg \in o_K[[G_0]]$ is not equal to the matrix (ca_{ij}).

Notice that

$$G_0 \subset o_K[G_0]^\times \subset o_K[[G_0]]^\times$$

where $o_K[G_0]^\times$ and $o_K[[G_0]]^\times$ are the groups of units in $o_K[G_0]$ and $o_K[[G_0]]$, respectively.

Exercise 2.48 Prove that $1 + \varpi_K o_K[[G_0]] \subset o_K[[G_0]]^\times$.

Example 2.49 (Iwasawa Rings for $G_0 = \mathbf{G}(o_L)$) Foreshadowing what will be done in Sect. 5.5, let $\mathbf{G}$ be a $\mathbb{Z}$-group and $G_0 = \mathbf{G}(o_L)$. As mentioned in Example 2.37, $G_0 = \varprojlim_{n \in \mathbb{N}} G_0/G_n$, where G_n is the kernel of $G_0 \to \mathbf{G}(o_L/\mathfrak{p}_L^n)$. Then

$$o_K[[G_0]] = \varprojlim_{n \in \mathbb{N} \cup \{0\}} o_K[G_0/G_n] = \varprojlim_{n \in \mathbb{N}} o_K[G_0/G_n].$$

2.3.2 Another Projective Limit Realization of $o_K[[G_0]]$

Notice that

$$o_K[[G_0]] = \varprojlim_{n \in N} o_K[G_0/G_n] = \varprojlim_{n \in N}(\varprojlim_{m \in \mathbb{N}} o_K/\mathfrak{p}_K^m)[G_0/G_n].$$

How we can combine the two projective limits on the right hand side into a single projective limit is described in the following proposition.

Proposition 2.50 *Suppose that* $m : N \to \mathbb{N} \cup \{0, \infty\}$ *is a function satisfying*

(i) $m(n) \leq m(n')$ *whenever* $n < n'$, *and*
(ii) *For any chain* $n_1 < n_2 < \cdots < n_i < \cdots$ *in* N, *we have* $\lim_{i \to \infty} m(n_i) = \infty$.

Then

$$o_K[[G_0]] \cong \varprojlim_{n \in N} o_K/\mathfrak{p}_K^{m(n)}[G_0/G_n].$$

Here, $\mathfrak{p}_K^0 = o_K$ *and* $\mathfrak{p}_K^\infty = 0$.

We allow $m(n) = 0$ because of our applications in Chap. 7 (see $m(\chi, n)$ defined on page 157). For the proof, we need the following lemma.

Lemma 2.51 *For* $n \geq n'$, *define* $\chi_{n,n'} : \mathfrak{p}_K^{m(n)}[G_0/G_n] \to \mathfrak{p}_K^{m(n')}[G_0/G_{n'}]$ *by*

$$\chi_{n,n'}(a_1 g_1 G_n + \cdots + a_s g_s G_n) = a_1 g_1 G_{n'} + \cdots + a_s g_s G_{n'}.$$

Then $(\mathfrak{p}_K^{m(n)}[G_0/G_n], \chi_{n,n'})_N$ *is an inverse system of topological* o_K-*modules, and*

$$\varprojlim_{n \in N} \mathfrak{p}_K^{m(n)}[G_0/G_n] = 0.$$

Proof The maps $\chi_{n,n'}$, over all $n \geq n'$, are compatible, so $(\mathfrak{p}_K^{m(n')}[G_0/G_{n'}], \chi_{n,n'})_N$ is an inverse system of topological rings.

Let $Y = \varprojlim_{n \in N} \mathfrak{p}_K^{m(n)}[G_0/G_n]$ and $X = \prod_n \mathfrak{p}_K^{m(n)}[G_0/G_n]$. Then $Y \subset X$. Take a nonzero $\mu \in X$. We claim that $\mu \notin Y$. Assume, on the contrary, that $\mu \in Y$. Write $\mu = (\mu_n)_{n \in N}$. There exists n_0 such that $\mu_{n_0} \neq 0$. It follows that there exists $\ell \in \mathbb{N}$ such that $\mu_{n_0} \notin \mathfrak{p}_K^\ell[G_0/G_{n_0}]$. By the properties of $m(n)$, there exists $n_1 \in N$ such that $m(n_1) \geq \ell$. Take $n_2 \geq n_0, n_1$. Then $\mu_{n_2} \in \mathfrak{p}_K^\ell[G_0/G_{n_2}]$. It follows

$$\mu_{n_0} = \chi_{n_2,n_0}(\mu_{n_2}) \in \mathfrak{p}_K^\ell[G_0/G_{n_0}],$$

a contradiction. Hence, $\mu \notin Y$, thus proving $Y = 0$. $\qquad\square$

Proof of Proposition 2.50 For every $n \in N$, we have the following exact sequence

$$0 \to \mathfrak{p}_K^{m(n)}[G_0/G_n] \to o_K[G_0/G_n] \to o_K/\mathfrak{p}_K^{m(n)}[G_0/G_n] \to 0.$$

We want to consider the corresponding inverse systems. For $n \geq n'$, we have the canonical projections

$$\varphi_{n,n'} : o_K[G_0/G_n] \to o_K[G_0/G_{n'}]$$

and

$$\psi_{n,n'} : o_K/\mathfrak{p}_K^{m(n)}[G_0/G_n] \to o_K/\mathfrak{p}_K^{m(n')}[G_0/G_{n'}].$$

We also have the homomorphism

$$\chi_{n,n'} : \mathfrak{p}_K^{m(n)}[G_0/G_n] \to \mathfrak{p}_K^{m(n')}[G_0/G_{n'}],$$

as in Lemma 2.51. Since all the maps are natural, it is easy to see that the following diagram is commutative, for all $n \geq n'$

$$
\begin{array}{ccccccccc}
0 \to & \mathfrak{p}_K^{m(n)}[G_0/G_n] & \to & o_K[G_0/G_n] & \to & o_K/\mathfrak{p}_K^{m(n)}[G_0/G_n] & \to 0 \\
& \downarrow & & \downarrow & & \downarrow & \\
0 \to & \mathfrak{p}_K^{m(n')}[G_0/G_{n'}] & \to & o_K[G_0/G_{n'}] & \to & o_K/\mathfrak{p}_K^{m(n')}[G_0/G_{n'}] & \to 0
\end{array}
$$

Hence, we have an exact sequence of morphisms of inverse systems

$$0 \to (\mathfrak{p}_K^{m(n)}[G_0/G_n])_N \to (o_K[G_0/G_n])_N \to (o_K/\mathfrak{p}_K^{m(n)}[G_0/G_n])_N \to 0.$$

In the projective limit, using Proposition 2.26 and Lemma 2.51, we get the exact sequence of topological rings

$$0 \to 0 \to o_K[[G_0]] \to \varprojlim_{n \in \mathcal{N}} o_K/\mathfrak{p}_K^{m(n)}[G_0/G_n] \to 0.$$

It follows $o_K[[G_0]] \cong \varprojlim_{n \in \mathcal{N}} o_K/\mathfrak{p}_K^{m(n)}[G_0/G_n]$. $\qquad\qquad\qquad\qquad\qquad\square$

Corollary 2.52 *If $G_0 \cong \varprojlim_{n \in \mathbb{N}} G_0/G_n$, then*

$$o_K[[G_0]] \cong \varprojlim_{n \in \mathbb{N}} o_K/\mathfrak{p}_K^n[G_0/G_n].$$

Proof Define $m : \mathbb{N} \to \mathbb{N}$ by $m(n) = n$ and apply Proposition 2.50. $\qquad\square$

2.3.3 Some Properties of Iwasawa Algebras

We conclude Chap. 2 with remarks on zero divisors, the augmentation map, and the Iwasawa algebra of a subgroup. Some additional important properties are listed in Proposition 5.42.

Zero Divisors

Suppose $g \in G_0$ is a torsion element, so $g^n = 1$ for some $n \in \mathbb{N}$. Then in $o_K[G_0] \subset o_K[[G_0]]$

$$(1 - g)(1 + g + g^2 + \cdots + g^{n-1}) = 0,$$

so $1 - g$ is a zero divisor.

Example 2.53 Let e be the identity element in $G_0 = GL_2(o_L)$ and $f = \begin{pmatrix} -1 & 0 \\ 0 & -1 \end{pmatrix}$. Notice that $f \neq -e$ in $o_K[[G_0]]$ (see the notation introduced on page 29). Then $1 + f = e + f \neq 0 \in o_K[[G_0]]$. We have

$$(1 + f)(1 - f) = 1 - f^2 = e - e = 0,$$

so $1 + f$ and $1 - f$ are zero divisors in $o_K[[G_0]]$.

For description of torsion elements in reductive groups, see [73]. If G_0 is a compact Lie group, then $\mathbb{Z}_p[[G_0]]$ has no zero divisors if and only if G_0 is torsion free [2, Theorem 4.3].

Augmentation Map

For the projective limit of topological rings, the projection maps are continuous ring homomorphisms. This was supposedly stated and proved in the solution to Exercise 2.23. Then for

$$o_K[[G_0]] = \varprojlim_{n \in N} o_K[G_0/G_n] = \varprojlim_{N \in \mathcal{N}(G_0)} o_K[G_0/N],$$

the projection maps $\varphi_n : o_K[[G_0]] \to o_K[G_0/G_n]$ are continuous ring homomorphisms. In particular, for $N = G_0$, we obtain the **augmentation map**

$$\text{aug} : o_K[[G_0]] \to o_K.$$

The augmentation map can be computed easily for $\mu \in o_K[G_0]$: if $\mu = a_1 g_1 + \cdots + a_s g_s$, then

$$\text{aug}(\mu) = \text{aug}(a_1 g_1 + \cdots + a_s g_s) = a_1 + \cdots + a_s.$$

If $\mathfrak{m}(G_0) = J_{1,0}(G_0)$ as in Lemma 2.43, then

$$\mathfrak{m}(G_0) = \{\mu \in o_K[[G_0]] \mid \text{aug}(\mu) \in \mathfrak{p}_K\}.$$

We know that $\mathfrak{m}(G_0)$ is a maximal ideal in $o_K[[G_0]]$.

Iwasawa Algebra of a Subgroup

Proposition 2.54 *Let G_0 be a profinite group and H a compact open subgroup of G_0.*

(i) The index of H in G_0 is finite.
(ii) $o_K[[H]]$ is a closed subalgebra of $o_K[[G_0]]$.
(iii) Let $\{g_1, \ldots, g_s\}$ be a set of coset representatives of G_0/H. Then

$$o_K[[G_0]] = g_1 o_K[[H]] \oplus \cdots \oplus g_s o_K[[H]].$$

(iv) Let $\{g_1, \ldots, g_s\}$ be a set of coset representatives of $H \backslash G_0$. Then

$$o_K[[G_0]] = o_K[[H]]g_1 \oplus \cdots \oplus o_K[[H]]g_s.$$

Proof

(i) The set of cosets gH is an open cover of G_0. By compactness, it has a finite subcover. Since the cosets are disjoint, it follows that H has a finite number of cosets in G, that is, $|G_0 : H|$ is finite.

(ii) Denote by $\mathcal{N}(H, G_0)$ the set of open normal subgroups of G_0 contained in H. From Proposition 2.32 and Lemma 2.15, we have $o_K[[H]] = \varprojlim_{N \in \mathcal{N}(H,G_0)} o_K[H/N]$ and

$$o_K[[G_0]] = \varprojlim_{N \in \mathcal{N}(G_0)} o_K[G_o/N] = \varprojlim_{N \in \mathcal{N}(H,G_0)} o_K[G_o/N].$$

Then the compatible embeddings $o_K[H/N] \hookrightarrow o_K[G_0/N]$, for $N \in \mathcal{N}(H, G_0)$, induce in the projective limit the continuous embedding

$$o_K[[H]] \hookrightarrow o_K[[G_0]].$$

Since $o_K[[H]]$ is compact, it is closed in $o_K[[G_0]]$.

(iii) The decompositions

$$o_K[G_o/N] = g_1 o_K[H/N] \oplus \cdots \oplus g_s o_K[H/N],$$

for $N \in \mathcal{N}(H, G_0)$, are compatible with respect to the natural projections. Passing to the projective limit gives us the statement.

$\square$

Chapter 3
Distributions

In this chapter, K is a finite extension of $\mathbb{Q}_p$ and G_0 is a profinite group

$$G_0 \cong \varprojlim_{n \in \mathcal{N}} G_0/G_n,$$

where $\mathcal{N}$ is a directed poset and $\{G_n \mid n \in \mathcal{N}\}$ is a family of open normal subgroups of G_0 such that $G_n \leq G_m$ for $n \geq m$.

The Iwasawa algebra $o_K[[G_0]]$ can be identified with the unit ball in the space of continuous distributions on G_0. Before explaining this identification, we need some background material on locally convex K-vector spaces.

3.1 Locally Convex Vector Spaces

In this section, we review basic definitions and properties of locally convex vector spaces. A detailed development of the theory of locally convex vector spaces can be found in Schneider's book [60].

Definition 3.1 Let V be a K-vector space.

(i) A subset $A \subset V$ is called **convex** if either A is empty or is of the form $A = v + A_0$ for some vector v and some o_K-submodule $A_0 \subset V$.
(ii) A **lattice** in V is an o_K-submodule $\mathcal{L}$ which satisfies the condition that for any $v \in V$ there is a nonzero scalar $a \in K^\times$ such that $av \in \mathcal{L}$.

© The Author(s), under exclusive license to Springer Nature Switzerland AG 2022
D. Ban, *p-adic Banach Space Representations*, Lecture Notes
in Mathematics 2325, https://doi.org/10.1007/978-3-031-22684-7_3

Let $(V, \| \ \|)$ be a normed K-vector space. For $v \in V$ and a real number $\epsilon > 0$,[1] we define the **closed ball** (or simply a **ball**) of radius ϵ centered at v

$$B_\epsilon(v) = \{w \in V \mid \|w - v\| \leq \epsilon\}$$

and the **open ball** $B_\epsilon^-(v) = \{w \in V \mid \|w - v\| < \epsilon\}$. It is a property of ultrametric spaces that every closed ball $B_\epsilon(v)$ is both open and closed (see Proposition A.3 in Appendix A). Similarly, every open ball $B_\epsilon^-(v)$ is both open and closed.

Exercise 3.2

(i) Prove that $B_\epsilon(0)$ and $B_\epsilon^-(0)$ are open lattices in V.
(ii) Prove that $B_\epsilon(v)$ and $B_\epsilon^-(v)$ are convex subsets.

Definition 3.3 Let V be a K-vector space. Let $(\mathcal{L}_j)_{j \in J}$ be a nonempty family of lattices in V such that

(lc1)   for any $j \in J$ and $a \in K^\times$, there exists a $k \in J$ such that $\mathcal{L}_k \subseteq a\mathcal{L}_j$, and
(lc2)   for any two $i, j \in J$ there exists a $k \in J$ such that $\mathcal{L}_k \subseteq \mathcal{L}_i \cap \mathcal{L}_j$.

Then the convex subsets of the form $v + \mathcal{L}_j$ for $v \in V$ and $j \in J$ form a basis of the topology on V called the **locally convex topology** on V defined by the family $(\mathcal{L}_j)_{j \in J}$.

Definition 3.4 A **locally convex K-vector space** is a K-vector space equipped with a locally convex topology.

Exercise 3.5 Let V be a locally convex K-vector space. Prove that the addition $V \times V \to V$ and scalar multiplication $K \times V \to V$ are continuous maps.

If $(V, \| \ \|)$ is a normed K-vector space, then the balls centered at 0 satisfy conditions (lc1) and (lc2). It follows that the convex subsets of the form $v + B_\epsilon(0)$, for $v \in V$ and $\epsilon \in \mathbb{R}^+$, form a basis of the locally convex topology on V. Since

$$v + B_\epsilon(0) = B_\epsilon(v),$$

this topology is equal to the topology on V induced by the norm $\| \ \|$. We will always consider the normed space V equipped with the locally convex topology described above.

Hence, any normed vector space is locally convex. In general, if V is a locally convex K-vector space, then the topology on V can be defined by a family of seminorms [60, Proposition 4.4].

Definition 3.6 Let V be a locally convex K-vector space. A subset $B \subset V$ is said to be **bounded** if for any open lattice $\mathcal{L} \subseteq V$ there is an $a \in K$ such that $B \subseteq a\mathcal{L}$.

---

[1] Yes, real.

For normed vector spaces, this definition coincides with the classical definition of boundedness, as we can see from the following exercise.

Exercise 3.7 Let $(V, \| \ \|)$ be a normed K-vector space. A subset $B \subset V$ is bounded if and only if the set $\|B\| = \{\|x\| \mid x \in B\}$ is bounded in $\mathbb{R}$.

3.1.1 Banach Spaces

The concept of completeness of a locally convex K-vector space V can be defined using Cauchy nets, as in [60, §7]. If V is metrizable, this is equivalent to the standard definition: V is complete if and only if every Cauchy sequence in V is convergent [60, Remark 7.2].

A Banach space is usually defined as a complete normed vector space. However, following the approach from [60], we do not consider the norm to be part of the structure.

Definition 3.8 A K-**Banach space** is a complete locally convex vector space whose topology can be defined by a norm.

3.1.2 Continuous Linear Operators

Suppose that V and W are locally convex K-vector spaces. We denote by

$$\mathcal{L}(V, W)$$

the space of all continuous linear maps $f : V \to W$. In particular, if $W = K$, we denote the space $\mathcal{L}(V, W)$ by V' and call it the **dual space** or **continuous dual** of V.

If V and W are normed K-vector spaces, for any K-linear map $f \in \mathrm{Hom}_K(V, W)$, we define the **operator norm**

$$\|f\| = \inf\{c \in [0, \infty) \mid \|f(v)\| \le c\|v\| \text{ for all } v \in V\}, \tag{3.1}$$

where the infimum of the empty set is ∞. If $V \ne 0$, then

$$\|f\| = \sup\{\frac{\|f(v)\|}{\|v\|} \mid v \in V, \|v\| \ne 0\}.$$

Exercise 3.9 Suppose that V and W are normed K-vector spaces, and $f \in \mathrm{Hom}_K(V, W)$. Prove that f is continuous if and only if $\|f\| < \infty$.

Notice that for every $f \in \mathcal{L}(V, W)$ and $v \in V$, we have

$$\|f(v)\| \leq \|f\| \, \|v\|.$$

Exercise 3.10 Suppose that V and W are normed K-vector spaces.

(a) Prove that Eq. (3.1) defines a nonarchimedean norm on $\mathcal{L}(V, W)$.
(b) If W is a Banach space, prove that $\mathcal{L}(V, W)$ is also Banach. *(See [60], Prop. 3.3.)*

Remark 3.11 The formula

$$\|f\|_0 = \sup\{\|f(v)\| \mid v \in V, \|v\| \leq 1\}$$

defines a norm on $\mathcal{L}(V, W)$ which is equivalent to the operator norm. Contrary to the real or complex case, $\|f\|_0$ and $\|f\|$ may not be equal. For instance, take $W = \mathbb{Q}_3$ with the 3-adic absolute value as the norm and also $V = \mathbb{Q}_3$, but with $\|v\| = 2|v|_3$. Let $f = \mathrm{id}_{\mathbb{Q}_3} : V \to W$. Then $\|f\| = 1/2$, but $\|f\|_0 = 1/3$.

Set $\|V\| = \{\|v\| \mid v \in V\}$ and $|K| = \{|a| \mid a \in K\}$.

Lemma 3.12 *If V and W are normed K-vector spaces such that $\|V\| = |K|$, then the operator norm on $\mathcal{L}(V, W)$ can be computed by the formula*

$$\|f\| = \sup\{\|f(v)\| \mid v \in V, \|v\| \leq 1\} = \sup\{\|f(v)\| \mid v \in V, \|v\| = 1\}.$$

Proof Since $\|V\| = |K|$, for every nonzero $v \in V$, there exists $a \in K$ such that $\|av\| = 1$. $\qquad\square$

From Exercise 3.10 we see that, for normed vector spaces V and W, $\mathcal{L}(V, W)$ carries the locally convex topology induced by the operator norm. As we will see below, this is just one among many. We review here a general method for defining different locally convex topologies on $\mathcal{L}(V, W)$. Details can be found in [60, §6].

Suppose that V and W are locally convex K-vector spaces. For a bounded subset $B \subset V$ and an open lattice $M \subset W$, we define

$$\mathcal{L}(B, M) = \{f \in \mathcal{L}(V, W) \mid f(B) \subseteq M\}.$$

This is a lattice in $\mathcal{L}(V, W)$. If W is a normed space, we write $\mathcal{L}(B, \epsilon)$ for $\mathcal{L}(B, B_\epsilon^-(0))$, i.e.,

$$\mathcal{L}(B, \epsilon) = \{f \in \mathcal{L}(V, W) \mid \|f(v)\| < \epsilon \text{ for all } v \in B\}.$$

Let now $\mathcal{B}$ be a fixed family of bounded subsets of V which is closed under finite unions. Then

$$\{\mathcal{L}(B, M) \mid B \in \mathcal{B}, M \text{ an open lattice in } W\}$$

is a family of lattices satisfying conditions (lc1) and (lc2) from Definition 3.3 (Exercise 3.13). It therefore defines a locally convex topology on $\mathcal{L}(V, W)$, called the $\mathcal{B}$-topology on $\mathcal{L}(V, W)$. We denote by $\mathcal{L}_{\mathcal{B}}(V, W)$ the space $\mathcal{L}(V, W)$ equipped with this topology.

Exercise 3.13 With notation as above,

(a) Prove that $\mathcal{L}(B, M)$ is a lattice in $\mathcal{L}(V, W)$.
(b) Prove that $\{\mathcal{L}(B, M) \mid B \in \mathcal{B}, M$ an open lattice in $W\}$ is a family of lattices satisfying conditions (lc1) and (lc2) from Definition 3.3

Exercise 3.14 Prove that for a bounded subset $B \subset V$, an open lattice $M \subset W$, and a nonzero $a \in K$, $\mathcal{L}(aB, M) = \mathcal{L}(B, a^{-1}M)$.

Important examples of $\mathcal{B}$-topologies on $\mathcal{L}(V, W)$ are the weak topology and the strong topology defined below.

Example 3.15 Let $\mathcal{B}$ be the family of all finite subsets of V. We denote $\mathcal{L}_{\mathcal{B}}(V, W)$ by

$$\mathcal{L}_s(V, W).$$

We call the corresponding $\mathcal{B}$-topology the **weak topology** or the **topology of pointwise convergence**. In particular, we write

$$V_s'$$

for the dual space V' equipped with the weak topology. In literature, this topology on V' is also called the weak* topology (pronounced "weak star topology"), with "star" referring to the dual space, sometimes denoted by V^*.

Exercise 3.16 For $v \in V$, we define the evaluation map $\mathrm{ev}_v : \mathcal{L}(V, W) \to W$ by

$$\mathrm{ev}_v : f \mapsto f(v), \quad f \in \mathcal{L}(V, W).$$

Prove that the weak topology is the weakest topology on $\mathcal{L}(V, W)$ making all evaluations $\mathrm{ev}_v : \mathcal{L}(V, W) \to W$ continuous.

Example 3.17 Let $\mathcal{B}$ be the family of all bounded subsets of V. We write

$$\mathcal{L}_b(V, W)$$

for $\mathcal{L}_{\mathcal{B}}(V, W)$. The corresponding $\mathcal{B}$-topology is called the **strong topology** or the **topology of bounded convergence**. In particular, we write

$$V_b'$$

for the dual space V' equipped with the strong topology.

Exercise 3.18 Suppose that V and W are normed vector spaces. Prove that the strong topology is same as the topology on $\mathcal{L}(V, W)$ induced by the operator norm.

3.1.3 Examples of Banach Spaces

For the basic example of K^n, see Example A.10 in Appendix A.

Banach Space of Bounded Functions

Let X be a nonempty set. Define

$$\ell^\infty(X) = \{f : X \to K \mid f \text{ bounded}\}.$$

Then $\ell^\infty(X)$ is a K-vector space with pointwise addition and scalar multiplication. The formula

$$\|f\|_\infty = \sup_{x \in X} |f(x)|$$

defines a norm on $\ell^\infty(X)$ which gives it the structure of a Banach space.

Let $c_0(X)$ be the subspace of $\ell^\infty(X)$ consisting of all $f \in \ell^\infty(X)$ such that for any $\epsilon > 0$ there exists only finitely many elements $x \in X$ such that $|f(x)| \geq \epsilon$. Then $c_0(X)$ is a closed subspace of $\ell^\infty(X)$ and hence $c_0(X)$ is also a Banach space. From [60, §3], we have

Lemma 3.19 *Let X be a nonempty set. Then the dual of $c_0(X)$ is isomorphic to $\ell^\infty(X)$.*

We write ℓ^∞ and c_0 for $\ell^\infty(\mathbb{N})$ and $c_0(\mathbb{N})$, respectively. Then ℓ^∞ is the space of bounded sequences and

$$c_0 = \{(a_1, a_2, \ldots) \in \ell^\infty \mid \lim_{n \to \infty} a_n = 0\}$$

is the space of null sequences.

Continuous Functions on G_0

Let $C(G_0, K)$ be the space of continuous K-valued functions on G_0. We equip $C(G_0, K)$ with the Banach space topology induced by the **sup norm**

$$\|f\| = \sup_{g \in G_0} |f(g)|,$$

which is by compactness of G_0 equal to $\max_{g \in G_0} |f(g)|$.

Exercise 3.20 Denote by $C^\infty(G_0, K)$ the subspace of $C(G_0, K)$ consisting of smooth (i.e., locally constant) functions. Prove that $C^\infty(G_0, K)$ is dense in $C(G_0, K)$.

Mahler Expansion

Let $G_0 = \mathbb{Z}_p$ and consider $C(\mathbb{Z}_p, K)$. For $x \in \mathbb{Z}_p$ and $n \geq 0$, define

$$\binom{x}{n} = \frac{x(x-1)\cdots(x-(n-1))}{n!}.$$

Theorem 3.21 (Mahler Expansion)

(i) *Let $f \in C(\mathbb{Z}_p, K)$. Then there exists a unique null sequence $(a_1, a_2, \ldots)$ in K such that*

$$f(x) = \sum_{n=1}^{\infty} a_n \binom{x}{n}, \quad x \in \mathbb{Z}_p.$$

The series converges uniformly and $\|f\| = \max_n |a_n|$.

(ii) *If $(a_1, a_2, \ldots)$ is a null sequence in K, then $x \mapsto \sum_{n=1}^{\infty} a_n \binom{x}{n}$ defines a continuous function $\mathbb{Z}_p \to K$.*

Proof Schikhof [59], Theorem 51.1. □

Corollary 3.22 *The dual of $C(\mathbb{Z}_p, K)$ is isomorphic to the space of bounded sequences.*

Proof By Mahler expansion, $C(\mathbb{Z}_p, K) \cong c_0$, and Lemma 3.19 implies $C(\mathbb{Z}_p, K) \cong \ell^\infty$. □

3.1.4 Double Duals of a Banach Space

In this section, we discuss the double duals of V. We assume that V is a K-Banach space—for simplicity of exposition, and also because our applications of the double duals will be in the duality theory for Banach spaces (Sect. 4.3).

The double duals of V depend on the topology we put on V'. Suppose that the topology on $V'_\mathcal{B}$ is finer than the topology on V'_s. It is also coarser than the topology on V_b'. Hence, for $f \in \mathrm{Hom}_K(V', K)$, we have

$$f \text{ continuous on } V_s' \quad \Rightarrow \quad f \text{ continuous on } V_\mathcal{B}',$$
$$f \text{ continuous on } V_\mathcal{B}' \quad \Rightarrow \quad f \text{ continuous on } V_b'.$$

It follows

$$(V_s')' \subseteq (V_\mathcal{B}')' \subseteq (V_b')'. \tag{3.2}$$

For $v \in V$, we have the evaluation map $\mathrm{ev}_v : V' \to K$ given by $\mathrm{ev}_v(\ell) = \ell(v)$, for all $\ell \in V'$. By Exercise 3.16, $\mathrm{ev}_v : V_s' \to K$ is continuous, so $\mathrm{ev}_v \in (V_s')'$. Then Eq. (3.2) tells us that ev_v belongs to $(V_\mathcal{B}')'$.

The linear map

$$\varepsilon : V \to (V_\mathcal{B}')'$$

given by $\varepsilon(v) = \mathrm{ev}_v$ is called a **duality map**. We would like to select $\mathcal{B}$ and equip $(V_\mathcal{B}')'$ with a locally convex topology so that $\varepsilon : V \to (V_\mathcal{B}')'$ is a topological isomorphism. The two obvious candidates, $(V_s')'_s$ and $(V_b')'_b$, usually do not give topological isomorphisms.

Definition 3.23 A K-Banach space V is called **reflexive** if the duality map $\varepsilon : V \to (V_b')'_b$ is a topological isomorphism.

Proposition 3.24 *Let V be a K-Banach space.*

(i) *The duality map*

$$\varepsilon : V \to (V_s')'_s$$

given by $\varepsilon(v) = \mathrm{ev}_v$ is a continuous bijection.

(ii) *The duality map*

$$\varepsilon : V \to (V_b')'_b$$

given by $\varepsilon(v) = \mathrm{ev}_v$ induces a topological isomorphism between V and $\mathrm{im}\,\varepsilon$.

(iii) *V is reflexive if and only if it is finite dimensional.*

Proof All statements can be found in [60], where they are proved in a more general context. (i) follows from Proposition 9.7, (ii) from Lemma 9.9, and (iii) from Proposition 11.1. $\qquad\square$

Since $\varepsilon : V \to (V_s')'_s$ is a continuous bijection, we can refine the topology on $(V_s')'_s$ so that we obtain a homeomorphism $\varepsilon : V \to (V_s')'$. It turns out that the strong topology on $(V_s')'$ works well.

Proposition 3.25 *The duality map*

$$\varepsilon : V \to (V_s')_b'$$

given by $\varepsilon(v) = \mathrm{ev}_v$ *is a topological isomorphism.*

Proof A locally convex vector space V is called barrelled if every closed lattice in V is open. Since our space V is Banach, it is barrelled (see Example 2 on page 35 in [60]). Then the proposition follows from Corollary 13.8 in [60].   □

3.2 Distributions

As introduced earlier, $C(G_0, K)$ is the Banach space of continuous K-valued functions on G_0. Let $D^c(G_0, K)$ be the continuous dual of $C(G_0, K)$. We have the canonical pairing $\langle \, , \, \rangle : D^c(G_0, K) \times C(G_0, K) \to K$ given by

$$\langle \mu, h \rangle = \mu(h).$$

From Exercise 3.10, we know that $D^c(G_0, K)$ is a Banach space with respect to the operator norm. This topology is equal to the strong topology (Exercise 3.18). Another important locally convex topology on $D^c(G_0, K)$ is of course the weak topology. In the theory of p-adic Banach space representations, another locally convex topology plays a prominent role. It is called the bounded weak topology and it is defined in Sect. 3.3. It coincides with the weak topology on the bounded sets. The weak topology on the unit ball in $D^c(G_0, K)$ is studied below.

3.2.1 The Weak Topology on $D^c(G_0, o_K)$

In this section, we restrict our attention to distributions which are o_K-valued on the o_K-valued functions. More specifically, with $C(G_0, o_K)$ denoting the o_K-module of all continuous functions $f : G_0 \to o_K$, define

$$D^c(G_0, o_K) = \{\mu \in D^c(G_0, K) \mid \langle \mu, f \rangle \in o_K \text{ for all } f \in C(G_0, o_K)\}.$$

Notice that $D^c(G_0, o_K)$ is the unit ball in $D^c(G_0, K)$, with respect to the operator norm. We equip $D^c(G_0, o_K)$ with the subspace topology coming from the weak topology on $D^c(G_0, K)$. We call this topology the weak topology on $D^c(G_0, o_K)$.

To describe a fundamental system of open neighborhoods of zero in $D^c(G_0, o_K)$, we first recall that the ideals $\mathfrak{p}_K^m$, $m \in \mathbb{N}$, form such a system in K. For a finite subset

$F \subset C(G_0, K)$ and $m \in \mathbb{N}$, we write $\mathcal{L}_0(F, m)$ for $\mathcal{L}(F, \mathfrak{p}_K^m) \cap D^c(G_0, o_K)$. Hence,

$$\mathcal{L}_0(F, m) = \{\mu \in D^c(G_0, o_K) \mid \langle \mu, f \rangle \in \mathfrak{p}_K^m \text{ for all } f \in F\}.$$

The o_K-modules $\mathcal{L}_0(F, m)$ form a fundamental system of open neighborhoods of zero in $D^c(G_0, o_K)$.

Lemma 3.26 *The weak topology on $D^c(G_0, o_K)$ has a fundamental system of open neighborhoods of zero consisting of the lattices*

$$\mathcal{L}_0(\{1_{U_1}, \ldots, 1_{U_s}\}, m),$$

where $U_1, \ldots, U_s$ are disjoint compact open subset of G_0 and $m \in \mathbb{N}$.

Proof Take an open lattice $\mathcal{L}_0(\{f\}, m)$, for some $f \in C(G_0, o_K)$ and $m \in \mathbb{N}$. By continuity of f, for every $g \in G_0$ there exists a compact open neighborhood U_g such that

$$f(u) - f(g) \in \mathfrak{p}_K^m, \quad \forall u \in U_g.$$

By compactness of G_0, there is a finite subset $\{g_1, \ldots, g_s\}$ of G_0 such that the sets U_{g_i}, for $i = 1, \ldots, s$, cover G_0. We can further reduce this to a disjoint cover by compact open subsets $U_i \subset U_{g_i}$. Set $a_i = f(g_i) \in o_K$. We claim that

$$\mathcal{L}_0(\{a_1 1_{U_1}, \ldots, a_s 1_{U_s}\}, m) \subseteq \mathcal{L}_0(\{f\}, m).$$

To prove the claim, take $\mu \in \mathcal{L}_0(\{a_1 1_{U_1}, \ldots, a_s 1_{U_s}\}, m)$. We can write $f = f_1 + \cdots + f_s$, where $f_i = f|_{U_i}$. Then

$$\langle \mu, f \rangle = \langle \mu, f_1 \rangle + \cdots + \langle \mu, f_s \rangle.$$

Fix $i \in \{1, \ldots, s\}$. By continuity of μ, there exists $\delta > 0$ such that

$$\|f_i - h\| < \delta \quad \Rightarrow \quad \langle \mu, f_i - h \rangle \in \mathfrak{p}_K^m,$$

for any $h \in C(G_0, o_K)$. Repeating a similar process as above, we can find a disjoint cover of U_i by compact open subsets U_{ij} and points $g_{ij} \in U_{ij}$ such that the function

$$h_i = b_{i1} 1_{U_{i1}} + \cdots + b_{ir} 1_{U_{ir}}$$

satisfies $b_{ij} = f(g_{ij})$ and $\|f_i - h_i\| < \delta$. Since the sets U_{ij} are disjoint, we have $1_{U_i} = 1_{U_{i1}} + \cdots + 1_{U_{ir}}$. Note that $g_{ij} \in U_{ij} \subset U_i$ and hence

$$b_{ij} - a_i = f(g_{ij}) - f(g_i) \in \mathfrak{p}_K^m.$$

Then

$$\langle \mu, h_i - a_i 1_{U_i} \rangle = \langle \mu, (b_{i1} 1_{U_{i1}} + \cdots + b_{ir} 1_{U_{ir}}) - a_i (1_{U_{i1}} + \cdots + 1_{U_{ir}}) \rangle$$

$$= \sum_{j=1}^{r} (b_{ij} - a_i) \langle \mu, 1_{U_{ij}} \rangle \in \mathfrak{p}_K^m,$$

(3.3)

because $\mu \in D^c(G_0, o_K)$, so $\langle \mu, 1_{U_{ij}} \rangle \in o_K$. It follows

$$\langle \mu, f \rangle = \langle \mu, f - \sum_i a_i 1_{U_i} \rangle + \sum_i \langle \mu, a_i 1_{U_i} \rangle$$

$$= \langle \mu, f - \sum_i h_i + \sum_i h_i - \sum_i a_i 1_{U_i} \rangle + \sum_i \langle \mu, a_i 1_{U_i} \rangle$$

$$= \sum_i \langle \mu, f_i - h_i \rangle + \sum_i \langle \mu, h_i - a_i 1_{U_i} \rangle + \sum_i \langle \mu, a_i 1_{U_i} \rangle.$$

Then $\sum_i \langle \mu, f_i - h_i \rangle \in \mathfrak{p}_K^m$ because $\|f_i - h_i\| < \delta$, Eq. (3.3) gives $\sum_i \langle \mu, h_i - a_i 1_{U_i} \rangle \in \mathfrak{p}_K^m$, and the assumption $\mu \in \mathcal{L}_0(\{a_1 1_{U_1}, \ldots, a_s 1_{U_s}\}, m)$ implies that $\sum_i \langle \mu, a_i 1_{U_i} \rangle \in \mathfrak{p}_K^m$. Hence, $\langle \mu, f \rangle \in \mathfrak{p}_K^m$, proving the claim. Since $a_i \in o_K$, for all i, we have

$$\mathcal{L}_0(\{1_{U_1}, \ldots, 1_{U_s}\}, m) \subseteq \mathcal{L}_0(\{a_1 1_{U_1}, \ldots, a_s 1_{U_s}\}, m) \subseteq \mathcal{L}_0(\{f\}, m).$$

Next, suppose that we have another element $f' \in C(G_0, o_K)$. Then there exist disjoint compact open subsets $U'_1, \ldots, U'_r$ of G_0 such that $\mathcal{L}_0(\{1_{U'_1}, \ldots, 1_{U'_r}\}, m) \subseteq \mathcal{L}_0(\{f'\}, m)$. Define $V_{ij} = U_i \cap U'_j$. Then

$$\mathcal{L}_0(\{1_{V_{ij}} \mid i = 1, \ldots, s, \ j = 1, \ldots, r\}, m) \subset \mathcal{L}_0(\{f\}, m) \cap \mathcal{L}_0(\{f'\}, m).$$

It follows that for any finite set $F \subset C(G_0, o_K)$ and any $m \in \mathbb{N}$ there exist disjoint compact open subsets $U_1, \ldots, U_r$ of G_0 such that $\mathcal{L}_0(\{1_{U_1}, \ldots, 1_{U_s}\}, m) \subseteq \mathcal{L}_0(F, m)$.

The final result is obtained by scaling. Namely, if F is a finite subset of $C(G_0, K)$, there exists a nonzero $a \in o_K$ such that $aF \subset C(G_0, o_K)$. For $m \in \mathbb{N}$, take $\ell \in \mathbb{N}$ such that $\mathfrak{p}_K^\ell \subset a\mathfrak{p}_K^m$. Then there exist disjoint compact open subsets $U_1, \ldots, U_r$ of G_0 such that

$$\mathcal{L}_0(\{1_{U_1}, \ldots, 1_{U_s}\}, \ell) \subseteq \mathcal{L}_0(aF, \ell) \subseteq \mathcal{L}_0(F, m),$$

finishing the proof. □

Corollary 3.27 Write $G_0 \cong \varprojlim_{n \in \mathbb{N}} G_0/G_n$, as on page 35. The weak topology on $D^c(G_0, o_K)$ has a fundamental system of open neighborhoods of zero consisting of *the lattices*

$$\mathcal{L}_0(\{1_{gG_n} \mid gG_n \in G_0/G_n\}, m),$$

where $m \in \mathbb{N}$, $n \in \mathcal{N}$.

Proof Take a lattice $\mathcal{L}_0(\{1_{U_1}, \ldots, 1_{U_s}\}, m)$, where $U_1, \ldots, U_s$ are disjoint compact open subset of G_0 and $m \in \mathbb{N}$. We can decompose each U_i as a disjoint union of cosets of G_n, for some $n \in \mathcal{N}$ (Exercise 3.28). Then

$$\mathcal{L}_0(\{1_{gG_n} \mid gG_n \in G_0/G_n\}, m) \subseteq \mathcal{L}_0(\{1_{U_1}, \ldots, 1_{U_s}\}, m)$$

and the statement follows from Lemma 3.26.   □

Exercise 3.28 Suppose that $U_1, \ldots, U_s$ are disjoint compact open subset of G_0. Prove that there exists $n \in \mathcal{N}$ such that each U_i can be written as a disjoint union of cosets of G_n.

3.2.2 Distributions and Iwasawa Rings

For $g \in G_0$, we denote by δ_g the corresponding **Dirac distribution**. This is the distribution in $D^c(G_0, o_K) \subset D^c(G_0, K)$ defined by $\delta_g(f) = f(g)$, for all $f \in C(G_0, K)$. Let

$$D^{Dir}(G_0, o_K)$$

denote the o_K-linear span in $D^c(G_0, o_K)$ of all Dirac distributions. This is an o_K-submodule of $D^c(G_0, o_K)$. Any element of $D^{Dir}(G_0, o_K)$ is a finite o_K-linear combination of Dirac distributions.

Lemma 3.29 *Let $n \in \mathcal{N}$. Define*

$$\mathcal{D}_n = \{\mu \in D^c(G_0, o_K) \mid \langle \mu, 1_{gG_n} \rangle = 0 \text{ for all } gG_n \in G_0/G_n\}.$$

Then

(i) $\mathcal{D}_n = \bigcap_{m \in \mathbb{N}} \mathcal{L}_0(\{1_{gG_n} \mid gG_n \in G_0/G_n\}, m)$.
(ii) For every $\mu \in D^c(G_0, o_K)$ there exists $\nu \in D^{Dir}(G_0, o_K)$ such that $\mu - \nu \in \mathcal{D}_n$.
(iii) If $\ell \geq n$, then $\mathcal{D}_\ell \subseteq \mathcal{D}_n$.

Proof Assertions (i) and (iii) are obvious. For (ii), take $\mu \in D^c(G_0, o_K)$. Let $\{g_1, \ldots, g_r\}$ be a set of coset representatives of G_0/G_n. Define

$$\nu = \langle \mu, 1_{g_1 G_n} \rangle \delta_{g_1} + \cdots + \langle \mu, 1_{g_r G_n} \rangle \delta_{g_r}.$$

This is an element of $D^{Dir}(G_0, o_K)$. For every $j \in \{1, \ldots, r\}$, we have

$$\langle \nu, 1_{g_j G_n} \rangle = \langle \mu, 1_{g_1 G_n} \rangle \delta_{g_1}(1_{g_j G_n}) + \cdots + \langle \mu, 1_{g_r G_n} \rangle \delta_{g_r}(1_{g_j G_n}) = \langle \mu, 1_{g_j G_n} \rangle.$$

It follows $\mu - \nu \in \mathcal{D}_n$. $\qquad\square$

Corollary 3.30 $D^{Dir}(G_0, o_K)$ *is dense in* $D^c(G_0, o_K)$*, where* $D^c(G_0, o_K)$ *is equipped with the weak topology.*

Proof Take an arbitrary neighborhood U of μ. By Corollary 3.27, U contains

$$\mu + \mathcal{L}_0(\{1_{g G_n} \mid g G_n \in G_0/G_n\}, m),$$

for some $m \in \mathbb{N}$, $n \in \mathcal{N}$. By Lemma 3.29, there exists $\nu \in D^{Dir}(G_0, o_K)$ such that $\mu - \nu \in \mathcal{D}_n \subset \mathcal{L}_0(\{1_{g G_n} \mid g G_n \in G_0/G_n\}, m)$. $\qquad\square$

Proposition 3.31 *We consider* $o_K[G_0]$ *equipped with the subspace topology coming from* $o_K[[G_0]]$ *and* $D^{Dir}(G_0, o_K)$ *equipped with the subspace topology coming from the weak topology on* $D^c(G_0, o_K)$. *The* o_K*-linear map* $o_K[G_0] \to D^{Dir}(G_0, o_K)$ *given by*

$$g \mapsto \delta_g, \quad g \in G_0,$$

is a topological isomorphism.

Proof Define $\varphi : o_K[G_0] \to D^{Dir}(G_0, o_K)$ by

$$\varphi(a_1 g_1 + \cdots + a_s g_s) = a_1 \delta_{g_1} + \cdots + a_s \delta_{g_s}.$$

This is an o_K-linear map and clearly a bijection. We denote the inverse of φ by ψ. Then $\psi(\delta_g) = g$.

To show that φ is a homeomorphism, we first recall that $o_K[[G_0]]$ has a fundamental system of open neighborhoods of zero consisting of the two-sided ideals $J_{m,n}$ (Lemma 2.43). We claim that

$$\varphi(J_{m,n} \cap o_K[G_0]) = \mathcal{L}_0(\{1_{g G_n} \mid g G_n \in G_0/G_n\}, m) \cap D^{Dir}(G_0, o_K). \qquad (3.4)$$

To prove the claim, take $\eta \in J_{m,n} \cap o_K[G_0]$. As an element of $o_K[G_0]$, η is a finite sum $\eta = \sum_{i=1}^{t} c_i h_i$ for some $h_i \in G_0$, $c_i \in o_K$. Let $\{g_1, \ldots, g_r\}$ be a set of coset representatives of G_0/G_n. Putting together h_i's belonging to the same coset of G_0/G_n, we can write η as

$$\eta = \sum_{j=1}^{r} \sum_{h_i \in g_j G_n} c_i h_i.$$

Then the projection η_n of η to $o_K[G_0/G_n]$ is

$$\eta_n = \sum_{j=1}^{r} \left(\sum_{h_i \in g_j G_n} c_i \right) g_j G_n.$$

The condition $\eta \in J_{m,n}$ implies $\sum_{h_i \in g_j G_n} c_i \in \mathfrak{p}_K^m$ for all j. For any $j \in \{1, \ldots, r\}$, we have

$$\langle \varphi(\eta), 1_{g_j G_n} \rangle = \langle \sum_{j=1}^{r} \sum_{h_i \in g_j G_n} c_i \delta_{h_i} , 1_{g_j G_n} \rangle = \sum_{h_i \in g_j G_n} c_i \in \mathfrak{p}_K^m.$$

Hence, $\varphi(\eta) \in \mathcal{L}_0(\{1_{g_1 G_n}, \ldots, 1_{g_r G_n}\}, m) \cap D^{Dir}(G_0, o_K)$.

The arguments for the converse inclusion are similar. Take

$$\eta \in \mathcal{L}_0(\{1_{g_1 G_n}, \ldots, 1_{g_r G_n}\}, m) \cap D^{Dir}(G_0, o_K).$$

Then $\eta = \sum_{i=1}^{t} c_i \delta_{h_i}$, for some $h_i \in G_0$ and some coefficients $c_i \in o_K$, and we can write it as

$$\eta = \sum_{j=1}^{r} \sum_{h_i \in g_j G_n} c_i \delta_{h_i}.$$

For each $j \in \{1, \ldots, r\}$, we have

$$\langle \eta, 1_{g_j G_n} \rangle = \sum_{h_i \in g_j G_n} c_i \in \mathfrak{p}_K^m.$$

This implies $\psi(\eta) \in \varphi(J_{m,n} \cap o_K[G_0])$, proving the claim.

It follows that both φ and ψ are continuous at zero. By linearity, both φ and ψ are continuous, thus proving that φ is a topological isomorphism.   □

Theorem 3.32 *The map*

$$g \mapsto \delta_g, \quad g \in G_0,$$

extends o_K-linearly and by continuity to a topological isomorphism of o_K-modules

$$o_K[[G_0]] \cong D^c(G_0, o_K),$$

where $D^c(G_0, o_K)$ carries the weak topology.

Proof Let $\psi : D^{Dir}(G_0, o_K) \to o_K[G_0]$ be, as in the proof of Proposition 3.31, the topological isomorphism of o_K-modules given by $\psi(\delta_g) = g$. Then for $n \in \mathcal{N}$

we have the corresponding continuous map

$$\psi_n : D^{Dir}(G_0, o_K) \to o_K[G_0/G_n]$$

obtained by composing ψ with the canonical projection $o_K[G_0] \to o_K[G_0/G_n]$. Let $\mathcal{J}_n$ and $\mathcal{D}_n$ be as in Lemmas 2.45 and 3.29, respectively. Equation (3.4) implies that for all $n \in N$, $\psi(\mathcal{D}_n \cap D^{Dir}(G_0, o_K)) = \mathcal{J}_n \cap o_K[G_0]$ and

$$\psi_n(\mathcal{D}_n \cap D^{Dir}(G_0, o_K)) = 0.$$

Then we have a well-defined o_K-linear map

$$\Psi_n : D^c(G_0, o_K) \to o_K[G_0/G_n]$$

given by $\Psi_n(\mu) = \psi_n(\nu)$, where ν is any element of $D^{Dir}(G_0, o_K)$ such that $\mu - \nu \in \mathcal{D}_n$. By Lemma 3.29, such ν exists. The maps Ψ_n, $n \in N$, are continuous, compatible, and induce in the projective limit the continuous surjection

$$\Psi : D^c(G_0, o_K) \to \varprojlim_{n \in N} o_K[G_0/G_n] = o_K[[G_0]]$$

(see Corollary 2.20). Notice that $\Psi|_{D^{Dir}(G_0,o_K)} = \psi$. It is easy to show that Ψ is injective, so Ψ is a continuous bijection

We claim that $\Psi(\mathcal{L}_0(\{1_{gG_n} \mid gG_n \in G_0/G_n\}, m)) = J_{m,n}(G_0)$, for any $m \in \mathbb{N}$, $n \in N$. Take $\mu \in \mathcal{L}_0(\{1_{gG_n} \mid gG_n \in G_0/G_n\}, m)$. There exists $\nu \in D^{Dir}(G_0, o_K) \cap \mathcal{L}_0(\{1_{gG_n} \mid gG_n \in G_0/G_n\}, m)$ such that $\mu - \nu \in \mathcal{D}_n$. Then $\Psi(\nu) = \psi(\nu) \in J_{m,n}(G_0)$ and $\Psi(\mu - \nu) \in \mathcal{J}_n$ imply $\Psi(\mu) \in J_{m,n}(G_0)$. This proves one containment.

For the converse containment, take $\mu \notin \mathcal{L}_0(\{1_{gG_n} \mid gG_n \in G_0/G_n\}, m)$. Then for some coset gG_n we have $\langle \mu, 1_{gG_n} \rangle \notin \mathfrak{p}_K^m$. By continuity of Ψ, there exists a neighborhood U of μ such that $\Psi(\eta) - \Psi(\mu) \in J_{m,n}(G_0)$ for all $\eta \in U$. Applying Lemma 3.29 once again, we can find $\nu \in D^{Dir}(G_0, o_K) \cap U$ such that $\mu - \nu \in \mathcal{D}_n$. Then $\langle \nu, 1_{gG_n} \rangle = \langle \mu, 1_{gG_n} \rangle \notin \mathfrak{p}_K^m$. It follows $\Psi(\nu) = \psi(\nu) \notin J_{m,n}(G_0)$ and hence $\Psi(\mu) \notin J_{m,n}(G_0)$.

It follows that Ψ is open, and hence a homeomorphism. $\square$

Corollary 3.33 $D^c(G_0, o_K)$ *is compact in* $D^c(G_0, K)$, *in the weak topology.*

Proof Since $o_K[[G_0]]$ is compact, Theorem 3.32 implies that $D^c(G_0, o_K)$ is compact as well. $\square$

3.2.3   The Canonical Pairing

It follows from Theorem 3.32 that we can identify $o_K[[G_0]]$ with $D^c(G_0, o_K)$ by identifying $g \in G_0$ with the Dirac distribution δ_g. Then from $\langle\,,\,\rangle : D^c(G_0, o_K) \times C(G_0, o_K) \to o_K$ we obtain the canonical pairing

$$\langle\,,\,\rangle : o_K[[G_0]] \times C(G_0, o_K) \to o_K.$$

We can describe the pairing explicitly (see §12 in [67]). Let $\mu \in o_K[[G_0]]$ and $h \in C(G_0, o_K)$. Write $\mu = (\mu_n)_{n=1}^\infty$, where $\mu_n \in o_K[G_0/G_n]$. On the other hand, h can be uniformly approximated by a sequence $\{h_n\}_{n=1}^\infty$ of smooth functions such that h_n is right G_n-invariant. If $g_1 G_n = g_2 G_n$, then $\delta_{g_1}(h_n) = \delta_{g_2}(h_n)$. It follows that we have a well-defined pairing $\langle \mu_n, h_n \rangle$. More specifically, if $\{g_1, \ldots, g_s\}$ is a set of representatives of G_0/G_n, we can write

$$\mu_n = a_1 g_1 G_n + \cdots + a_s g_s G_n \quad \text{and} \quad h_n = b_1 1_{g_1 G_n} + \cdots + b_s 1_{g_s G_n},$$

where $a_i \in o_K$ and $b_i \in o_K$ for all i. Then

$$\langle \mu_n, h_n \rangle = a_1 b_1 + \cdots + a_s b_s. \tag{3.5}$$

It can be shown that $\{\langle \mu_n, h_n \rangle\}_{n=1}^\infty$ is a Cauchy sequence whose limit is independent of the choice of $\{h_n\}_{n=1}^\infty$ (Exercise 3.34). Then

$$\langle \mu, h \rangle = \lim_{n \to \infty} \langle \mu_n, h_n \rangle.$$

Observe that $h_n \in C(G_0, o_K)$, so we can apply the above formula to evaluate $\langle \mu, h_n \rangle$. It is easy to show that $\langle \mu, h_n \rangle = \langle \mu_n, h_n \rangle$.

Exercise 3.34 Let $\langle \mu_n, h_n \rangle$ be as in Eq. (3.5). Prove that $\{\langle \mu_n, h_n \rangle\}_{n=1}^\infty$ is a Cauchy sequence whose limit is independent of the choice of $\{h_n\}_{n=1}^\infty$.

3.3   The Bounded-Weak Topology

By a **norm-bounded subset** of V', we mean a subset $B \subset V'$ bounded with respect to the operator norm, which is equivalent to B being bounded in V'_b.

Definition 3.35 Let V be a normed K-vector space. The **bounded-weak topology** on V' is the finest locally convex topology on V' that coincides with the weak topology on norm-bounded subsets of V'. We denote by

$$V'_{bs}$$

the space V' equipped with the bounded-weak topology.

In the Schneider-Teitelbaum duality theory described in Sects. 4.3 and 4.4, the dual V' of a p-adic Banach space representation V carries the bounded weak topology. It is therefore important for us to understand this topology well, and we will devote the next two sections to its study, with Sect. 3.4 taking care of some properties of $K[[G_0]]$ needed in Chap. 4. More general properties of the bounded-weak topology can be found in [57], where this topology is called the *bounded weak star topology*.

The following lemma gives several criteria for determining if a lattice is open in V'_{bs}. The conditions are very similar, but it will be useful for us having them all stated.

Lemma 3.36 *Let V be a normed K-vector space and M the unit ball in V'. Then the following are equivalent for any lattice $\mathcal{L} \subseteq V'$:*

(i) *$\mathcal{L}$ is open in V'_{bs};*
(ii) *for every norm-bounded set $B \subset V'$, the set $\mathcal{L} \cap B$ is open in B, where B carries the weak topology;*
(iii) *for any $0 \neq c \in o_K$, the set $c\mathcal{L} \cap M$ is open in M, where M carries the weak topology;*
(iv) *for any $k \in \mathbb{N}$, the set $\mathcal{L} \cap \varpi_K^{-k} M$ is open in $\varpi_K^{-k} M$, where $\varpi_K^{-k} M$ carries the weak topology.*

Proof (i) $\Leftrightarrow$ (ii) by the definition of the bounded-weak topology.

(iii) $\Leftrightarrow$ (iv) follows from the property that scalar multiplication $K \times V' \to V'$ is continuous in any locally convex topology (Exercise 3.5), using the fact that a nonzero $c \in o_K$ can be written as $c = \varpi_K^k u$, for some $k \in \mathbb{N} \cup \{0\}$ and a unit element $u \in o_K^\times$.

To prove the equivalence (ii) $\Leftrightarrow$ (iv), assume first (ii). For any $k \in \mathbb{N}$, the set $\varpi_K^{-k} M$ is norm-bounded. Then (ii) implies that $\mathcal{L} \cap \varpi_K^{-k} M$ is open in $\varpi_K^{-k} M$.

Conversely, assume (iv). Let B be a norm-bounded subset of V'. Then $B \subset \varpi_K^{-k} M$, for some $k \in \mathbb{N}$. By (iv), $\mathcal{L} \cap \varpi_K^{-k} M$ is open in $\varpi_K^{-k} M$. Then $\mathcal{L} \cap B$ is open in B by the definition of subspace topology. $\qquad\square$

Lemma 3.37 *Let V be a normed K-vector space and M the unit ball in V'. Then the bounded-weak topology on V' is the finest locally convex topology such that the inclusion of M, with its weak topology, is continuous.*

Proof Let us denote by V'_{lc} (just in this proof) the space V' equipped with the finest locally convex topology such that the inclusion of M, with its weak topology, is continuous. Hence, V'_{lc} contains as many open lattices as possible. The only obstruction comes from the requirement that

$$\iota : M \hookrightarrow V'$$

is continuous. If $\mathcal{L}$ is an open lattice in V'_{lc}, then necessarily $\mathcal{L} \cap M$ is open in M. This condition, however, is not sufficient. Recall that from the condition (lc1) in Definition 3.3, if $\mathcal{L}$ is an open lattice in V'_{lc}, then $a\mathcal{L}$ is also open for any $a \in K^\times$.

Hence, $a\mathcal{L} \cap M$ must be open in M for any $a \in K^{\times}$. This reduces to the requirement that $c\mathcal{L} \cap M$ is open in M for any $0 \neq c \in o_K$. The statement then follows from Lemma 3.36. $\qquad\square$

Let us look again at the unit ball M, equipped with the weak topology, and the inclusion $\iota : M \hookrightarrow V'$. Recall that the **final topology** with respect to ι is the finest topology making ι continuous. Notice that the finest locally convex topology making ι continuous is not the final topology with respect to ι. The final topology is usually much finer, with many more open sets. The description of the bounded-weak topology in terms of the final topology is given below.

Lemma 3.38 *Let V be a normed K-vector space and M the unit ball in V'. For $k \in \mathbb{N}$, let $\varpi_E^{-k} M$ carry the weak topology, and let ι_k be the embedding*

$$\iota_k : \varpi_E^{-k} M \hookrightarrow V'.$$

Then the bounded-weak topology on V' is the final topology with respect to the family of maps $(\iota_k)_{k \in \mathbb{N}}$. In other words, $V'_{bs} = \bigcup_{k \in \mathbb{N}} \varpi_E^{-k} M$ is the topological union.

Proof Exercise. $\qquad\square$

If $\mathcal{L}$ is an open lattice in V'_s, then $\mathcal{L}$ is open in V'_{bs}, so the bounded-weak topology is finer than the weak topology. We will show below that it is strictly finer. Still, V'_s and V'_{bs} have same dual spaces, as we can see from the following lemma.

Lemma 3.39 *Let V be K-Banach space. Then*

$$(V'_s)' = (V'_{bs})'.$$

Proof Follows form Proposition 3.2 and Corollary 3.3 of [58]. (Also, see Corollary 2.2 (i) of [57].) For a comment on normpolar spaces, see Remark 4.21. $\qquad\square$

Corollary 3.40 *Let V be a K-Banach space. The duality map*

$$\varepsilon : V \to (V'_{bs})'_b$$

given by $\varepsilon(v) = \mathrm{ev}_v$ is a topological isomorphism.

Proof Follows from Lemma 3.39 and Proposition 3.25. $\qquad\square$

3.3.1 The Bounded-Weak Topology is Strictly Finer than the Weak Topology

In this section, we present an example showing that the bounded-weak topology is not same as the weak topology.

Let $K = \mathbb{Q}_p$. Let $V = c_0$ be the space of null sequences in $\mathbb{Q}_p$, that is

$$V = \{(a_n)_n \in \prod_{\mathbb{N}} \mathbb{Q}_p \mid \lim_{n \to \infty} a_n = 0\}.$$

The norm is given by $\|(a_n)_n\| = \max_n |a_n|$. We know from Lemma 3.19 that the dual of V is isomorphic to the space of bounded sequences ℓ^∞,

$$V' \cong \{(b_n)_n \in \prod_{\mathbb{N}} \mathbb{Q}_p \mid |b_n| \text{ bounded}\},$$

where the pairing $V' \times V \to \mathbb{Q}_p$ is given by $\langle (b_n)_n, (a_n)_n \rangle \mapsto \sum_n a_n b_n$. The operator norm on V' is then $\|(b_n)_n\| = \max_n |b_n|$, and the unit ball in V' is

$$M = \{(b_n)_n \in \prod_{n \in \mathbb{N}} \mathbb{Q}_p \mid |b_n| \leq 1 \; \forall n\} \cong \prod_{\mathbb{N}} \mathbb{Z}_p.$$

The Weak Topology on V'

For $i \in \mathbb{N}$, we define $\mathbf{e}_i = (\delta_{i,n})_n \in V$ and $\mathbf{f}_i = (\delta_{i,n})_n \in V'$, where $\delta_{i,n}$ is the Kronecker delta. We have

$$\mathcal{L}(\{\mathbf{e}_i\}, k) = \{\mathbf{b} \in V' \mid \langle \mathbf{e}_i, \mathbf{b} \rangle \in p^k \mathbb{Z}_p\}$$

$$= \{\mathbf{b} = (b_n)_n \in V' \mid b_i \in p^k \mathbb{Z}_p\} \subset \prod_{n < i} \mathbb{Q}_p \times p^k \mathbb{Z}_p \times \prod_{n > i} \mathbb{Q}_p.$$

For $i \in \mathbb{N}$ and $k \in \mathbb{Z}$, define

$$U_{i,k} = \mathcal{L}(\{\mathbf{e}_i\}, k).$$

Lemma 3.41 *Let A be a finite subset of V and $\ell \in \mathbb{Z}$. Then there exist at most finitely many $i \in \mathbb{N}$ such that $\mathcal{L}(A, \ell) \subset U_{i,k}$ for some $k \in \mathbb{Z}$.*

Proof We may assume that A is linearly independent. Set

$$I(A) = \{i \in \mathbb{N} \mid a_i \neq 0 \text{ for some } \mathbf{a} = (a_n)_n \in A\}.$$

For $j \notin I(A)$ we have $\langle \mathbf{a}, \mathbf{f}_j \rangle = 0$ for all $\mathbf{a} \in A$. It follows $c\mathbf{f}_j \in \mathcal{L}(A, \ell)$ for any $c \in \mathbb{Q}_p$. Hence, $\mathcal{L}(A, \ell) \not\subset U_{j,k}$ for any $j \notin I(A)$ and any $k \in \mathbb{Z}$. Thus, the statement is clearly true if $I(A)$ is finite.

Suppose $I(A)$ is infinite. Denote by $J(A)$ the set of indices $j \in I(A)$ such that A contains a multiple of $\mathbf{e}_j$. Fix $i \in I(A) \setminus J(A)$. We can select a finite set $\{i = i_1, i_2, \ldots, i_m\} \subset I(A)$ such that

$$\mathrm{rank}\{(a_{i_1}, a_{i_2}, \ldots, a_{i_m}) \mid \mathbf{a} = (a_n)_n \in A\} = \mathrm{rank}\{(a_{i_2}, \ldots, a_{i_m}) \mid \mathbf{a} = (a_n)_n \in A\}.$$

Then we can find $c_2, \ldots, c_m \in \mathbb{Q}_p$ such that

$$a_{i_1} + c_2 a_{i_2} + \cdots + c_m a_{i_m} = 0$$

for all $\mathbf{a} = (a_n)_n \in A$. Set $\mathbf{b} = \mathbf{f}_{i_1} + c_2 \mathbf{f}_{i_2} + \cdots + c_m \mathbf{f}_{i_m}$. Then $\langle \mathbf{a}, \mathbf{b} \rangle = 0$, for all $\mathbf{a} \in A$. It follows that $c\,\mathbf{b} \in \mathcal{L}(A, \ell)$ for any $c \in \mathbb{Q}_p$. Hence, $\mathcal{L}(A, \ell) \not\subset U_{i,k}$ for any $k \in \mathbb{Z}$.

Since we can repeat the process for any $i \in I(A) \setminus J(A)$ and $J(A)$ is finite, this completes the proof. $\qquad\qquad\qquad\qquad\qquad\qquad\qquad\qquad\qquad\qquad\qquad\qquad\qquad\qquad\qquad\square$

The Bounded-Weak Topology on V'

We consider the unit ball M with the weak topology. Similarly, for any $k \in \mathbb{N}$, we consider $p^{-k}M$ equipped with the weak topology. Then $U \subset V'_{bs}$ is open if and only if $U \cap p^{-k}M$ is open for all $k \geq 0$.

Define

$$\mathcal{L}_0 = M = \prod_{n \in \mathbb{N}} \mathbb{Z}_p, \quad \mathcal{L}_1 = \mathbb{Z}_p \times \prod_{n>1} p^{-1}\mathbb{Z}_p, \quad \mathcal{L}_2 = \mathbb{Z}_p \times p^{-1}\mathbb{Z}_p \times \prod_{n>2} p^{-2}\mathbb{Z}_p,$$

and for any $k \geq 0$

$$\mathcal{L}_k = \mathbb{Z}_p \times p^{-1}\mathbb{Z}_p \times \cdots \times p^{-k+1}\mathbb{Z}_p \times \prod_{n>k} p^{-k}\mathbb{Z}_p.$$

Set $\mathcal{L} = \bigcup_{k \geq 0} \mathcal{L}_k$. Note that for every $k \geq 0$,

$$\mathcal{L} \cap p^{-k}M = \mathcal{L}_k = \bigcap_{n=1}^{k} U_{n, -n+1} \cap p^{-k}M.$$

is open in $p^{-k}M$. It follows that $\mathcal{L}$ is open in V'_{bs}. On the other hand,

$$\mathcal{L} \subset \bigcap_{n \in \mathbb{N}} U_{n,-n+1}.$$

Lemma 3.41 implies that $\mathcal{L}$ is not open in V'_s.

3.4 Locally Convex Topology on $K[[G_0]]$

Recall that $D^c(G_0, K)$ is a Banach space with respect to the operator norm. The unit ball $D^c(G_0, o_K)$ is a lattice in $D^c(G_0, K)$ and hence

$$D^c(G_0, K) = \bigcup_{k \in \mathbb{N}} \varpi_K^{-k} D^c(G_0, o_K).$$

From Theorem 3.32, we know that $D^c(G_0, o_K)$ is isomorphic to the Iwasawa algebra $o_K[[G_0]]$, where $D^c(G_0, o_K)$ carries the weak topology and $o_K[[G_0]]$ carries the projective limit topology. Our next step is to define the Iwasawa algebra corresponding to $D^c(G_0, K)$. Since the o_K-module $o_K[[G_0]]$ is torsion-free, the map $\mu \mapsto 1 \otimes \mu$ gives us an inclusion $o_K[[G_0]] \hookrightarrow K \otimes_{o_K} o_K[[G_0]]$.

Definition 3.42 Let G_0 be a profinite group. Define

$$K[[G_0]] = K \otimes_{o_K} o_K[[G_0]],$$

equipped with the finest locally convex topology such that the inclusion $o_K[[G_0]] \hookrightarrow K[[G_0]]$ is continuous.

Notice that the definition of $K[[G_0]]$ specifies the topology. By Theorem 3.44 below, we can identify $K[[G_0]]$ with $D^c(G_0, K)$. Then, we could consider other topologies on $K[[G_0]]$, defined by their $D^c(G_0, K)$-counterparts. An important example can be found in Sect. 4.4.1. In this book, however, we always consider $K[[G_0]]$ with the topology as in Definition 3.42. In particular, Lemma 1.4 of [64] implies

Lemma 3.43 $K[[G_0]]$ is complete.

If $a \in K$ and $\mu \in o_K[[G_0]]$, we write simply $a\mu$ for the element $a \otimes \mu \in K[[G_0]]$.

Theorem 3.44 *The map*

$$g \mapsto \delta_g, \quad g \in G_0,$$

extends K-linearly and by continuity to a topological isomorphism of K-vector spaces

$$K[[G_0]] \cong D^c(G_0, K),$$

where $D^c(G_0, K)$ carries the bounded-weak topology. Thus, we can identify $K[[G_0]]$ and $D^c(G_0, K)$ by identifying $g \in G_0$ with the Dirac distribution δ_g.

Proof Let $\mu \in K[[G_0]]$. Then $\mu = a\eta$, for some $a \in K$ and $\eta \in o_K[[G_0]]$. If $a \in o_K$, then $\mu \in o_K[[G_0]]$. Otherwise, we can write $a = \varpi_K^{-k} u$, with $k \in \mathbb{N}$ and $u \in o_K^\times$. Then

$$\mu \in a o_K[[G_0]] = \varpi_K^{-k} o_K[[G_0]].$$

It follows

$$K[[G_0]] = \bigcup_{k \in \mathbb{N}} \varpi_K^{-k} o_K[[G_0]].$$

From Theorem 3.32, we know that $o_K[[G_0]] \cong D^c(G_0, o_K)$ as topological rings and o_K-modules, where the isomorphism is obtained by extending o_K-linearly and by continuity the map $g \mapsto \delta_g$. Then also

$$\varpi_K^{-k} o_K[[G_0]] \cong \varpi_K^{-k} D^c(G_0, o_K)$$

for any $k \in \mathbb{N}$. Lemma 3.36 implies that the map in question is an isomorphism of locally convex K-vector spaces.   □

3.4.1   The Canonical Pairing

The dual pairing $\langle\,,\,\rangle : o_K[[G_0]] \times C(G_0, o_K) \to o_K$ can be extended K-bilinearly to

$$\langle\,,\,\rangle : K[[G_0]] \times C(G_0, K) \to K. \tag{3.6}$$

Alternatively, we identify $K[[G_0]]$ with $D^c(G_0, K)$ as in Theorem 3.44 and use the pairing $\langle\,,\,\rangle : D^c(G_0, K) \times C(G_0, K) \to K$. The resulting pairing $\langle\,,\,\rangle : K[[G_0]] \times C(G_0, K) \to K$ is same as in (3.6).

If we fix $\mu \in K[[G_0]]$, then $f \mapsto \langle \mu, f \rangle$ defines a map

$$\langle \mu,\,\rangle : C(G_0, K) \to K.$$

If we identify μ with an element of $D^c(G_0, K)$, then $\langle \mu, f \rangle = \mu(f)$ and hence $\langle \mu, \rangle = \mu$. On the other hand, if we fix $f \in C(G_0, K)$, then $\mu \mapsto \langle \mu, f \rangle$ defines a map

$$\langle \, , f \rangle : K[[G_0]] \to K.$$

This is precisely the evaluation map $ev_f : V' \to K$, for the Banach space $V = C(G_0, K)$ and its dual $V' = K[[G_0]]$. The evaluation map is continuous if we equip V' with the weak topology, or any topology finer than the weak one. It follows that $\langle \, , f \rangle : K[[G_0]] \to K$ is continuous. Moreover, the map

$$f \mapsto \langle \, , f \rangle$$

is precisely the duality map $\epsilon : V \to (V'_{bs})'_b$ and it is a topological isomorphism by Corollary 3.40.

We can use integral notation and write

$$\langle \mu, f \rangle = \int_{G_0} f(x) d\mu(x),$$

as it is done in [67, §12]. More details about the integration pairing in the context of locally analytic distributions can be found in [65, Section 2].

This is maybe a good place to say something about p-adic integration and also about the great obstacle: nonexistence of a p-adic Haar measure on $\mathbb{Z}_p$ and on other p-adic groups.

3.4.2 p-adic Haar Measure

The problem with defining a K-valued Haar measure on a pro-p-group comes from the properties of p-adic absolute value. Let us look for instance at $\mathbb{Z}_p$. Suppose that we have a p-adic measure μ on $\mathbb{Z}_p$ which is translation invariant. If $\mu \neq 0$, we may scale it to get $\mu(\mathbb{Z}_p) = 1$. We can write $\mathbb{Z}_p$ as a disjoint union of cosets of $p\mathbb{Z}_p$,

$$\mathbb{Z}_p = \coprod_{a=0}^{p-1} (a + p\mathbb{Z}_p).$$

Then from $\mu(a + p\mathbb{Z}_p) = \mu(p\mathbb{Z}_p)$ and

$$\mu(\mathbb{Z}_p) = \sum_{a=0}^{p-1} \mu(a + p\mathbb{Z}_p)$$

we get $\mu(p\mathbb{Z}_p) = 1/p$. This looks reasonable, unless we think how "big" are the numbers 1 and $1/p$. These are p-adic numbers, with

$$|\mu(\mathbb{Z}_p)|_p = 1 \quad \text{and} \quad |\mu(p\mathbb{Z}_p)|_p = |1/p|_p = p.$$

Now, this does not look right, because the subset $p\mathbb{Z}_p$ is measured with a "bigger" number than $\mathbb{Z}_p$. We start rightfully suspecting that a p-adic Haar measure on $\mathbb{Z}_p$ does not exist. Before proving our suspicions, of course, we have to define a Haar measure, so that we know what it is that does not exist (see [59, 30.4] and [77, pages 248 and 305]).

Definition 3.45 Let G be a locally compact group. Denote by $\mathcal{B}(G)$ the set of all compact open subsets of G. A K-valued **left Haar measure** on G is a map $\mu :$ $\mathcal{B}(G) \to K$ satisfying

  (i) additivity: $\mu(A \cup B) = \mu(A) + \mu(B)$, for any two disjoint sets $A, B \in \mathcal{B}(G)$,
 (ii) translation invariance: $\mu(gA) = \mu(A)$ for any $A \in \mathcal{B}(G)$ and $g \in G$, and
(iii) boundedness: for every $A \in \mathcal{B}(G)$, the set $\{\mu(B) \mid B \in \mathcal{B}(G), B \subset A\}$ is bounded.

Proposition 3.46 (Nonexistence of a p-adic Haar Measure on $\mathbb{Z}_p$) *If μ is a K-valued Haar measure on $\mathbb{Z}_p$, then $\mu = 0$.*

Proof Let $\mu : \mathcal{B}(\mathbb{Z}_p) \to K$ be a K-valued Haar measure on $\mathbb{Z}_p$. Set $c = \sup\{|\mu(A)|_p \mid A \in \mathcal{B}(\mathbb{Z}_p)\}$. By the boundedness requirement, we have $c < \infty$. For any $n \in \mathbb{N}$, we can write $\mathbb{Z}_p$ as a disjoint union of cosets of $p^n\mathbb{Z}_p$,

$$\mathbb{Z}_p = \coprod_{a=0}^{p^n-1} (a + p^n\mathbb{Z}_p).$$

By additivity and translation invariance, we have $\mu(\mathbb{Z}_p) = p^n\mu(p^n\mathbb{Z}_p)$. Then $|\mu(\mathbb{Z}_p)|_p = p^{-n}|\mu(p^n\mathbb{Z}_p)|_p \leq p^{-n}c$. It follows

$$|\mu(\mathbb{Z}_p)|_p \leq \lim_{n\to\infty} p^{-n}c = 0.$$

Hence, $\mu(\mathbb{Z}_p) = 0$. ◻

Not only $\mathbb{Z}_p$, but also any of the groups $\mathbf{G}(L)$, where L is a finite extension of $\mathbb{Q}_p$ and $\mathbf{G}$ is a reductive L-group, do not posses a nontrivial K-valued Haar measure. More generally, Monna and Springer proved in [47] that a locally compact group G has a K-valued Haar measure if and only if it contains a *p-free* compact open subgroup H. Here, H being p-free means that H does not contain an open subgroup whose index is divisible by p.

3.4.3   The Ring Structure on $D^c(G_0, K)$

The ring structure on $o_K[[G_0]]$ induces the ring structure on $K[[G_0]]$, which can be transferred to $D^c(G_0, K)$ using identification from Theorem 3.44. We denote by $*$ the corresponding multiplication map

$$* : D^c(G_0, K) \times D^c(G_0, K) \to D^c(G_0, K)$$

and call it the **convolution product**. Notice that for $g, h \in G_0$ we have

$$\delta_{gh} = \delta_g * \delta_h.$$

There is also a direct approach for defining the convolution product on $D^c(G_0, K)$. Continuous distributions are contained in the ring of locally analytic distributions. For details about the convolution product of locally analytic distribution, built using the integration notation, see Section 2 in [65].

Recall that $o_K[[G_0]]$ is a topological ring, and therefore the multiplication $o_K[[G_0]] \times o_K[[G_0]] \to o_K[[G_0]]$ is continuous, where $o_K[[G_0]] \times o_K[[G_0]]$ carries the product topology.

Proposition 3.47 *Let G_0 be a profinite group. The multiplication*

$$K[[G_0]] \times K[[G_0]] \to K[[G_0]]$$

is separately continuous (meaning that the map $\eta \mapsto \mu\eta$ is continuous for fixed μ, and $\mu \mapsto \mu\eta$ is continuous for fixed η).

Proof Fix $\mu \in K[[G_0]]$. We will prove that the map $f_\mu : K[[G_0]] \to K[[G_0]]$ given by $f_\mu(\eta) = \mu\eta$ is continuous. Since $f_\mu(\eta + \xi) = f_\mu(\eta) + f_\mu(\xi)$, it is enough to prove continuity at zero. For that, take an open lattice $\mathcal{L}$ in $K[[G_0]]$. Notice that for every $m, n \in \mathbb{Z}$, the multiplication gives a continuous map

$$\varpi_K^{-m} o_K[[G_0]] \times \varpi_K^{-n} o_K[[G_0]] \to \varpi_K^{-(m+n)} o_K[[G_0]].$$

Let m be an integer such that $\mu \in \varpi_K^{-m} o_K[[G_0]]$. For $n \in \mathbb{N}$, let f_n be the restriction of f_μ to $\varpi_K^{-n} o_K[[G_0]]$. Then

$$f_n : \varpi_K^{-n} o_K[[G_0]] \to \varpi_K^{-(m+n)} o_K[[G_0]]$$

is continuous. The lattice $\mathcal{L} \cap \varpi_K^{-(m+n)} o_K[[G_0]]$ is open in $\varpi_K^{-(m+n)} o_K[[G_0]]$ and hence

$$\mathcal{L}_n := f_n^{-1}(\mathcal{L} \cap \varpi_K^{-(m+n)} o_K[[G_0]])$$

is an open lattice in $\varpi_K^{-n} o_K[[G_0]]$. We claim that

$$\mathcal{L}_n = f_\mu^{-1}(\mathcal{L}) \cap \varpi_K^{-n} o_K[[G_0]].$$

To prove the claim, take $\xi \in f_\mu^{-1}(\mathcal{L}) \cap \varpi_K^{-n} o_K[[G_0]]$. Then $f_n(\xi) = f_\mu(\xi) \in \mathcal{L}$, so $f_n(\xi) \in \mathcal{L}$ and $f_n(\xi) \in \varpi_K^{-(m+n)} o_K[[G_0]]$. It follows $\xi \in \mathcal{L}_n$. Conversely, if $\xi \in \mathcal{L}_n$, then $\xi \in f_\mu^{-1}(\mathcal{L})$ and $\xi \in \varpi_K^{-n} o_K[[G_0]]$, proving the claim. Define

$$\mathcal{L}' = \bigcup_{n \in \mathbb{N}} \mathcal{L}_n.$$

Then $\mathcal{L}' = f_\mu^{-1}(\mathcal{L})$. The lattice $\mathcal{L}'$ is open in $K[[G_0]]$ because $\mathcal{L}' \cap \varpi_K^{-n} o_K[[G_0]] = \mathcal{L}_n$ is open in $\varpi_K^{-n} o_K[[G_0]]$ for every $n \in \mathbb{N}$. $\qquad\square$

Corollary 3.48 *With $D^c(G_0, K)$ carrying the bounded-weak topology, the convolution product*

$$* : D^c(G_0, K) \times D^c(G_0, K) \to D^c(G_0, K)$$

is separately continuous.

A Big Projective Limit

It is tempting to try to express $K[[G_0]]$ as a projective limit. We describe below something that at first may appear as a natural candidate, but we will recognize immediately that it is not.

For $m, n \in \mathbb{N}$, $m < n$, we have the natural projection

$$\varphi_{n,m} : K[G_0/G_n] \to K[G_0/G_m].$$

Then $(K[G_0/G_n], \varphi_{n,m})_{\mathbb{N}}$ is an inverse system, and we can define

$$\varprojlim_{n \in \mathbb{N}} K[G_0/G_n].$$

Exercise 3.49

(a) Let $\mu \in K[[G_0]]$. Then there exist $a \in K$ and $\eta \in o_K[[G_0]]$ such that $\mu = a\eta$. Write $\eta = (\eta_n)_{n \in \mathbb{N}}$, $\eta_n \in o_K[G_0/G_n]$. Define

$$\mu_n = a\eta_n \in K[G_0/G_n].$$

Prove that the definition of μ_n does not depend on the choice of a and η.

(b) From (a), we have a well-defined map $\varphi_n : K[[G_0]] \to K[G_0/G_n]$ given by $\varphi_n(\mu) = \mu_n$. Prove that the maps φ_n, $n \in \mathbb{N}$, are compatible and continuous, and hence induce in the projective limit the continuous map

$$\varphi : K[[G_0]] \to \varprojlim_{n \in \mathbb{N}} K[G_0/G_n].$$

Prove that φ is injective.

(c) Prove that $\varprojlim_{n \in \mathbb{N}} K[G_0/G_n]$ is bigger than $K[[G_0]]$ by constructing an element of $\varprojlim_{n \in \mathbb{N}} K[G_0/G_n]$ that is not in $K[[G_0]]$.

(d) After solving (c), we realize that $\varprojlim_{n \in \mathbb{N}} K[G_0/G_n]$ is much bigger than $K[[G_0]]$. Describe how they are related. Discuss topological properties.

Chapter 4
Banach Space Representations

Throughout the book, K and L are finite extensions of $\mathbb{Q}_p$.

For the most of the chapter, the group G_0 is a profinite group. From Sect. 4.4 on, we require that G_0 is a compact p-adic Lie group. Then we have the fundamental result of Lazard that the Iwasawa algebra $o_K[[G_0]]$ is noetherian. This property, together with Schikhof's duality, is a basis of the theory of admissible Banach space representations by Schneider and Teitelbaum [64].

In this chapter, we present results of [64]. Our writing is complementary to their. We introduce definitions and prove some basic properties, and then refer to [64] for the proofs of main results. These results are for compact Lie groups and can be simply extended to arbitrary Lie groups, which is done in Section 4.4.2.

4.1 p-adic Lie Groups

We refer to [63] for the theory p-adic Lie groups. We will not use this theory directly—the dependance on it is hidden in the proof that $o_K[[G_0]]$ is noetherian. For convenience of the reader, we review briefly the notion of p-adic Lie groups.

Let L be a finite extension of $\mathbb{Q}_p$. A **Lie group** (over L) is a manifold (over L) which also carries the structure of a group such the multiplication map $G \times G \to G$ is locally analytic (see [63, Section 13]). Such groups are also called **locally L-analytic groups**.

A p-**adic Lie group** is a Lie group over $\mathbb{Q}_p$. In Part II, we will consider a split reductive $\mathbb{Z}$-group $\mathbf{G}$. Then $G = \mathbf{G}(L)$ and $G_0 = \mathbf{G}(o_L)$ are locally L-analytic groups. If $L \neq \mathbb{Q}_p$, we can use the restriction of scalars to realize G and G_0 as locally $\mathbb{Q}_p$-analytic groups. Hence, G and G_0 are p-adic Lie groups, with G_0 in addition being compact, and the results of Sect. 4.4 apply to them.

© The Author(s), under exclusive license to Springer Nature Switzerland AG 2022   63
D. Ban, *p-adic Banach Space Representations*, Lecture Notes
in Mathematics 2325, https://doi.org/10.1007/978-3-031-22684-7_4

4.2 Linear Operators on Banach Spaces

Recall that we define a K-Banach space as a complete locally convex vector space whose topology can be defined by a norm (see Definition 3.8). We denote by

$$\mathrm{Ban}(K)$$

the category of all K-Banach spaces with morphisms being all continuous K-linear maps.

In our definition of a Banach space, the norm is not considered to be part of the structure. Hence, if V is a K-Banach space, there exists a norm $\|\ \|$ on V such that the balls

$$B_\epsilon(v_0) = \{v \in V \mid \|v - v_0\| < \epsilon\}$$

for $v_0 \in V$ and $\epsilon \in \mathbb{R}^+$, form a base of the topology. However, this norm is not fixed. One of the advantages of this approach is that we can change the norm, as in the following lemma.

Lemma 4.1 *Let V be a K-Banach space. Then there exists a norm $\|\ \|$ on V inducing the Banach space topology such that $\|V\| \subset |K|$.*

Proof Let $\|\ \|$ be a norm inducing the Banach space topology on V. For $v \in V$, define

$$\|v\|' = \inf\{r \in |K| \mid r \geq \|v\|\}.$$

Then $\|\ \|'$ is another norm on V. For every nonzero $v \in V$, we have $|\varpi_K| \leq \|v\|/\|v\|' \leq 1$. It follows that the norms $\|\ \|$ and $\|\ \|'$ are equivalent, so $\|\ \|'$ also induces the original Banach space topology on V. This norm satisfies $\|V\|' \subset |K|$.
$\square$

4.2.1 Spherically Complete Spaces

The concept of spherical completeness plays a fundamental role in the p-adic functional analysis. It is defined for ultrametric spaces. As explained in Appendix A, an ultrametric space is a metric space (X, d) such that the metric d satisfies the strong triangle inequality.

Definition 4.2 An ultrametric space (X, d) is said to be **spherically complete** if every nested sequence of balls in X has nonempty intersection.

Lemma 4.3 *Let (X, d) be an ultrametric space. The following conditions are equivalent*

(i) *X is spherically complete.*
(ii) *If $\mathcal{B}$ is a collection of balls such that no two elements of $\mathcal{B}$ are disjoint, then $\bigcap_{B \in \mathcal{B}} B \neq \emptyset$.*
(iii) *Every sequence of balls $B(x_1, \epsilon_1) \supset B(x_2, \epsilon_2) \supset \cdots$ for which $\epsilon_1 > \epsilon_2 > \cdots$ has nonempty intersection.*

Proof Exercise. □

Lemma 4.4 *If (X, d) is a complete ultrametric space and if every strictly decreasing sequence of values of d converges to 0, then (X, d) is spherically complete.*

Proof Let $B(x_1, \epsilon_1) \supset B(x_2, \epsilon_2) \supset \cdots$ be a sequence of balls in X such that $\epsilon_1 > \epsilon_2 > \cdots$ In every ball $B(x_n, \epsilon_n)$, we select $y_n \in B(x_n, \epsilon_n)$. The sequence $(y_n)_{n \in \mathbb{N}}$ is Cauchy, so it converges in X to $y = \lim_{n \to \infty} y_n$. Then $\{y\} = \bigcap_{n \in \mathbb{N}} B(x_n, \epsilon_n)$. □

Corollary 4.5

(i) *K is spherically complete.*
(ii) *If V is a K-Banach space, there exists a norm $\| \ \|$ on V such that V is spherically complete with respect to the metric induced by $\| \ \|$.*

Proof

(i) By Lemma A.8, K satisfies the requirements of Lemma 4.4 and it is therefore spherically complete.
(ii) By Lemma 4.1, there exists a norm $\| \ \|$ on V inducing the Banach space topology such that $\|V\| \subset |K|$. Then V is spherically complete by Lemma 4.4. □

As an example of a complete nonarchimedean field which is not spherically complete, we mention $\mathbb{C}_p$, the completion of $\overline{\mathbb{Q}}_p$, as defined in Appendix A.2.3. The proof that $\mathbb{C}_p$ is not spherically complete can be found in [60, §1].

4.2.2 Some Fundamental Theorems in Functional Analysis

We give several important theorems from functional analysis, for Banach spaces, as in [77]. More general versions can be found in [60].

In the proof, we use Baire spaces. A topological space X is called a **Baire space** if it satisfies the following condition. Given any countable collection $\{A_n\}$ of closed sets of X each of which has empty interior in X, their union $\bigcup_n A_n$ also has empty interior in X [48, page 295]. From the Baire category theorem, complete metric spaces are Baire spaces [48, Theorem 48.2].

We also need the notion of the sum of a series. It is defined in the standard way, as follows. Let V be a K-Banach space, and $v_1, v_2, \ldots$ a sequence in V. If the sequence of partial sums $s_n = \sum_{i=1}^{n} v_i$ converges to $s = \lim_{n \to \infty} s_n$, we say that the series $\sum_{n=1}^{\infty} v_n$ is convergent, with sum s, and we write

$$s = \sum_{n=1}^{\infty} v_n.$$

A wonderful thing in p-adic analysis is that the convergence of such series can be determined easily.

Lemma 4.6 *Let V be a K-Banach space, and $v_1, v_2, \ldots$ a sequence in V. The series $\sum_{n=1}^{\infty} v_n$ converges if and only if*

$$\lim_{n \to \infty} v_n = 0.$$

Proof Suppose $\lim_{n \to \infty} v_n = 0$. We will prove that the sequence of partial sums $s_1, s_2, \ldots$ is Cauchy. Let $\epsilon > 0$. Then there exists $N \in \mathbb{N}$ such that for every $n > N$ we have $\|v_n\| < \epsilon$. Then for $m > n > N$, using the strong triangle inequality, we get

$$\|s_m - s_n\| = \|\sum_{j=n+1}^{m} v_j\| \leq \max(\|v_{n+1}\|, \ldots, \|v_m\|) < \epsilon.$$

It follows that the sequence $(s_n)_n$ is Cauchy, hence convergent. The converse is obvious. $\square$

Theorem 4.7 (Closed Graph Theorem) *Let $f : V \to W$ be a K-linear map of K-Banach spaces. If its graph $\{(v, f(v)) \mid v \in V\}$ is a closed subset of $V \times W$, then f is continuous.*

Proof We will prove that there exists $\delta > 0$ such that for all $v \in B_\delta(0)$ we have $\|f(v)\| \leq 1$. By linearity, this implies that f is continuous.

Let A be the closure in V of the set $\{v \in V \mid \|f(v)\| \leq 1\}$. Then

$$V = \bigcup_{n \geq 0} \varpi_K^{-n} A$$

with each $\varpi_K^{-n} A$ closed. As a Banach space, V is also a Baire space, so one of the sets $\varpi_K^{-n} A$ has nonempty interior. Therefore, there exist $m \in \mathbb{N}$, $\epsilon > 0$, and $u \in V$ such that $B_\epsilon(u) \subset \varpi_K^{-m} A$. It follows

$$B_\epsilon(0) \subset u + \varpi_K^{-m} A \subset \varpi_K^{-m} A + \varpi_K^{-m} A = \varpi_K^{-m} A.$$

Set $\delta = |\varpi_K|^m \epsilon$. Then $B_\delta(0) \subset A$. Take any $v \in B_\delta(0)$. Then, there exists v_0 such that $\|f(v_0)\| \leq 1$ and $\|v - v_0\| \leq |\varpi_K| \delta$. Then $\|\varpi_K^{-1} v - \varpi_K^{-1} v_0\| \leq \delta$, and so

$\varpi_K^{-1} v - \varpi_K^{-1} v_0 \in B_\delta(0)$. Inductively, we can construct a sequence $v_0, v_1, \ldots$ such that

$$\| f(v_n) \| \leq 1,$$

$$\| \varpi_K^{-n-1} v - \varpi_K^{-n-1} v_0 - \varpi_K^{-n} v_1 - \cdots - \varpi_K^{-1} v_n \| \leq \delta,$$

for all n. Then $\| v - (v_0 + \varpi_K v_1 + \cdots + \varpi_K^n v_n) \| \leq |\varpi_K|^{n+1} \delta$, for all n. It follows

$$v = \sum_n \varpi_K^n v_n.$$

Since $\| f(v_n) \| \leq 1$, we have

$$\lim_{n \to \infty} \| \varpi_K^n f(v_n) \| \leq \lim_{n \to \infty} \| \varpi_K^n \| = 0.$$

Lemma 4.6 implies that $\sum_n \varpi_K^n f(v_n)$ converges. Since the graph of f is closed, it follows $f(v) = \sum_n \varpi_K^n f(v_n)$. Then $\| f(v) \| \leq 1$, completing the proof. □

Corollary 4.8 *If $f : V \to W$ is a continuous linear bijection of K-Banach spaces, then f is a homeomorphism.*

Proof Since the graph of f is closed in $V \times W$, and f is a bijection, it follows that the graph of f^{-1} is closed in $W \times V$. Hence, f^{-1} is continuous. □

Recall that a map $f : X \to Y$ between topological spaces is said to be **open** if it maps open sets to open sets.

Theorem 4.9 (Open Mapping Theorem) *Suppose that V and W are K-Banach spaces. Then every surjective continuous linear map $f : V \to W$ is open.*

Proof Let $U = \ker f$ and denote by F the induced continuous linear bijection $F : V/U \to W$. By Corollary 4.8, F is a homeomorphism. If O is an open set in V, we denote by $\bar{O}$ its image in V/U. Then $\bar{O}$ is open in V/U, and $F(\bar{O}) = f(O)$ is open in W. □

Theorem 4.10 (Uniform Boundedness Theorem (Banach-Steinhaus)) *Let V be a Banach space and W a normed vector space over K. Let $S \subseteq \mathcal{L}(V, W)$. If*

$$\sup_{f \in S} \| f(v) \| < \infty$$

for all $v \in V$, then

$$\sup_{f \in S} \| f \| < \infty.$$

Here, $\| f \|$ is the operator norm.

Proof Let $S \subseteq \mathcal{L}(V, W)$ be as in the statement of the theorem. For every $n \in \mathbb{N}$, the set

$$V_n = \{v \in V \mid \|f(v)\| \leq n \text{ for all } f \in S\}$$

is a closed o_K-submodule of V. By the assumption on S, $V = \bigcup_{n \in \mathbb{N}} V_n$. As mentioned before, V is a Baire space. It follows that one of the V_n has nonempty interior, so it must contain an open neighborhood of zero. Fix such an n and $\epsilon > 0$ such that $B_\epsilon(0) \subset V_n$.

Take an arbitrary $v \in V$. Let $m = m(v)$ be the least integer such that $|\varpi_K|^m \|v\| \leq \epsilon$. Then $|\varpi_K|^{m-1} \|v\| \geq \epsilon$, so

$$\epsilon |\varpi_K| \leq |\varpi_K|^m \|v\| \leq \epsilon.$$

The second inequality implies $\varpi_K^m v \in B_\epsilon(0)$, so $\|f(\varpi_K^m v)\| \leq n$ for all $f \in S$. From the first inequality, taking reciprocal, we get $|\varpi_K^{-m}| \leq \dfrac{1}{\epsilon |\varpi_K|} \|v\|$. Then for any $f \in S$

$$\|f(v)\| = \|\varpi_K^{-m} f(\varpi_K^m v)\| \leq |\varpi_K^{-m}| n \leq \frac{n}{\epsilon |\varpi_K|} \|v\|.$$

Notice that this shows $\|f(v)\| \leq \dfrac{n}{\epsilon |\varpi_K|} \|v\|$, for all $v \in V$ and therefore $\|f\| \leq \dfrac{n}{\epsilon |\varpi_K|}$. This holds for any $f \in S$ and consequently, $\sup_{f \in S} \|f\| \leq \dfrac{n}{\epsilon |\varpi_K|}$. $\qquad \square$

Corollary 4.11 *Suppose that* $f_1, f_2, \ldots$ *is a sequence in* $\mathcal{L}(V, W)$ *such that* $\lim_{n \to \infty} f_n(v)$ *exists for every* $v \in V$. *Define* $f : V \to W$ *by*

$$f(v) = \lim_{n \to \infty} f_n(v).$$

Then $f \in \mathcal{L}(V, W)$.

Proof The map f is clearly K-linear. We have to show that it is continuous. By the uniform boundedness theorem, there exists a number c such that $\|f_n\| \leq c$ for all $n \in \mathbb{N}$. Then $\|f(v)\| \leq c \|v\|$ for all $v \in V$. It follows that f is bounded, and hence it is continuous. $\qquad \square$

4.2.3 Banach Space Representations: Definition and Basic Properties

If V is a K-Banach space, we denote by $\mathrm{Aut}^c(V)$ the group of all continuous automorphisms of V. It is a subspace $\mathcal{L}(V, V)$, and it can carry a subspace topology.

In particular, we write $\mathrm{Aut}_s^c(V)$ when it is equipped with the weak topology coming from the embedding $\mathrm{Aut}^c(V) \subset \mathcal{L}_s(V, V)$

Lemma 4.12 *Let G be a locally profinite group and let V be a K-Banach space.*

(i) *Suppose that G acts on V by continuous linear automorphisms such that the map $G \times V \to V$ describing the action is continuous. Define*

$$\pi : G \to \mathrm{Aut}_s^c(V) \subset \mathcal{L}_s(V, V)$$

by $\pi(g)v = gv$. Then π is a continuous homomorphism.

(ii) *Conversely, suppose that we have a continuous homomorphism $\pi : G \to \mathrm{Aut}_s^c(V)$. Define the action of G on V by $gv = \pi(g)v$. Then the map $G \times V \to V$ describing the action is continuous.*

Proof Giving an action of G on V by continuous linear automorphisms is equivalent to giving a homomorphism $\pi : G \to \mathrm{Aut}^c(V)$. We have to prove continuity. For $g \in G$, denote $\pi(g)$ by f_g, so

$$f_g(v) = \pi(g)v = gv, \quad \text{for all } v \in V.$$

(i) This direction is the easy one, but let us write the details anyway.

Suppose $G \times V \to V$ is continuous. Take $g \in G$ and a neighborhood of f_g in $\mathcal{L}_s(V, V)$. Then it contains a neighborhood of the form

$$f_g + \mathcal{L}(\{v_1, \ldots, v_s\}, \epsilon),$$

for some $v_1, \ldots, v_s \in V$ and $\epsilon > 0$. By continuity of the action, for every $i \in \{1, \ldots, s\}$ there exists an open neighborhood U_i of g such that $\|hv_i - gv_i\| < \epsilon$, $\forall h \in U_i$. Let $U = \bigcap_{i=1}^s U_i$. This is an open neighborhood of g. For any $h \in U$,

$$\|f_h(v_i) - f_g(v_i)\| < \epsilon, \quad \forall i \in \{1, \ldots, s\}.$$

It follows $f_h - f_g \in \mathcal{L}(\{v_1, \ldots, v_s\}, \epsilon)$, and hence $f_h \in f_g + \mathcal{L}(\{v_1, \ldots, v_s\}, \epsilon)$. This proves that $\pi : G \to \mathcal{L}_s(V, V)$ given by $g \mapsto f_g$ is continuous.

(ii) Suppose $\pi : G \to \mathcal{L}_s(V, V)$ is continuous. For every $v \in V$, the evaluation map $\mathrm{ev}_v : \mathcal{L}_s(V, V) \to V$ given by $\mathrm{ev}_v(f) = f(v)$ is continuous (see Exercise 3.16). Since π and $\| \|$ are continuous, the map

$$\| \| \circ \mathrm{ev}_v \circ \pi : G \to \mathbb{R}_{\geq 0}$$

$$h \mapsto \|f_h(v)\|$$

is continuous. If U is a compact subset of G, then for every $v \in V$ the set

$$\{\|f_h(v)\| \mid h \in U\}$$

is bounded, because the continuous image of a compact set is compact.

Now, let $g \in G$ and $v_0 \in V$, and take an open neighborhood of gv_0. It contains an open ball $B(gv_0, \epsilon)$ for some $\epsilon > 0$. Let us consider $f_g + \mathcal{L}(\{v_0\}, \epsilon)$, which is a neighborhood of f_g in $\mathcal{L}_s(V, V)$. By continuity of π, there exists a compact open neighborhood U of g such that

$$f_h \in f_g + \mathcal{L}(\{v_0\}, \epsilon), \quad \forall h \in U.$$

As observed above, the set $\{\|f_h(v)\| \mid h \in U\}$ is bounded. The Banach-Steinhaus theorem (Theorem 4.10) implies that $\{\|f_h\| \mid h \in U\}$ is bounded by a constant $c > 0$. Then for every $h \in U$ and every $v \in B(v_0, \epsilon/c)$, we have

$$\|hv - gv_0\| = \|f_h(v) - f_h(v_0) + f_h(v_0) - f_g(v_0)\|$$
$$\leq \max\{\|f_h(v) - f_h(v_0)\|, \|f_h(v_0) - f_g(v_0)\|\}$$
$$\leq \max\{\|f_h\|\|v - v_0\|, \epsilon\} = \epsilon,$$

thus proving that $G \times V \to V$ is continuous.

□

Definition 4.13 Let G be a locally profinite group and V a K-Banach space. A **K-Banach space representation** of G on V is a continuous homomorphism

$$\pi : G \to \mathrm{Aut}_s^c(V) \subset \mathcal{L}_s(V, V),$$

or, equivalently, a G-action on V by continuous linear automorphisms such that the map $G \times V \to V$ describing the action is continuous. We denote this Banach space representation by (π, V), or just V, or just π.

We need some standard terminology from representation theory, such as intertwining operators, equivalence of representations, and subrepresentations. We introduce it below for Banach space representations, but similar definitions work for other types of representations, such as smooth representations considered in Chap. 6.

If (π, V) and (τ, W) are two K-Banach space representations of the same group G, we denote by

$$\mathrm{Hom}_G^c(\pi, \tau) \quad \text{or} \quad \mathrm{Hom}_G^c(V, W)$$

the space of all continuous K-linear maps $f : V \to W$ satisfying

$$f \circ \pi(g) = \tau(g) \circ f, \quad \text{for all } g \in G.$$

We call such maps **intertwining operators**. The condition above can be also written as

$$f(gv) = gf(v), \quad \text{for all } g \in G, v \in V,$$

and we say that f is G-**equivariant**. If $f \in \text{Hom}_G^c(\pi, \tau)$ is bijective, we know from Lemma 4.8 that f is a homeomorphism. If so, say that π and τ are **isomorphic** or **equivalent**, and we write $\pi \cong \tau$ or $V \cong W$.

We denote by

$$\text{Ban}_G(K)$$

the category of all K-Banach space representations of G with morphisms being all G-equivariant continuous linear maps.

Let (π, V) be a K-Banach space representation of G. A subspace U of V is said to be G-**invariant** (or G-**stable**) if

$$\pi(g) v \in U, \quad \forall g \in G, v \in U.$$

We also say that U is a **subrepresentation** of V. The representation (π, V) is called **topologically irreducible** if $V \neq 0$ and V has no G-invariant closed subspaces except 0 and itself.

Exercise 4.14 Suppose G_0 is compact. Then $V = C(G_0, K)$, with the sup norm, is a Banach space. There are two natural actions of G_0 on V:

(a) $g \in G_0$ acts on $f \in V$ by **left translation** L_g, where

$$L_g f(x) = f(g^{-1}x), \quad \forall x \in G_0.$$

(b) $g \in G_0$ acts on $f \in V$ by **right translation** R_g, where

$$R_g(x) = f(xg), \quad \forall x \in G_0.$$

Prove that each action defines a Banach space representation. Prove that these two representations are equivalent.

Exercise 4.15 Let $V = C(G_0, K)$ with the G_0-action as in the previous exercise (either (a) or (b)). Let $V^\infty = C^\infty(G_0, K)$ be the subspace of $C(G_0, K)$ consisting of smooth (i.e., locally constant) functions. Prove that V^∞ is G_0-invariant. Hence, it is a dense subrepresentation of V.

Exercise 4.16 Suppose V and W are two Banach space representation of G. Let $f \in \text{Hom}_G^c(V, W)$. Prove that $\ker f$ is a closed G-invariant subspace of V and that $\text{im } f$ is a G-invariant subspace of W.

Exercise 4.17 Let V be a K-Banach space representation of G. Suppose U is a closed G-invariant subspace of V. We consider the exact sequence of spaces

$$0 \to U \to V \to V/U \to 0.$$

The action of G on V induces the canonical action of G on U/V given by

$$g(v + U) = gv + U, \quad g \in G, v \in V.$$

Prove that this is a Banach representation of G on V/U. It is called a **quotient representation**.

Exercise 4.18 Let V be a K-Banach space representation of G. From Exercise 3.10, we know that V', equipped with the operator norm, is also a K-Banach space. We define the action of $g \in G$ on $\lambda \in V'$ by $g\lambda = {}^g\lambda$, where

$$^g\lambda(v) = \lambda(g^{-1}v), \quad \forall v \in V.$$

Prove that this defines a Banach space representation of G on V', called the **dual representation** of V.

Taking the dual $V \mapsto V'$ defines a functor $\mathrm{Ban}_G(K) \to \mathrm{Ban}_G(K)$. Of course, for a functor, we also have to define it on morphisms. This is done in a standard way. Given $\varphi \in \mathrm{Hom}_G^c(V, W)$ and $\lambda \in W'$, using the following commutative diagram

we define $\varphi'(\lambda) = \lambda \circ \varphi$. It is easy to show that $\varphi' \in \mathrm{Hom}_G^c(W', V')$ and that the map $\varphi \mapsto \varphi'$ satisfies the properties in the definition of a contravariant functor.

Exercise 4.19 Prove that $V \mapsto V'$ is a contravariant functor $\mathrm{Ban}_G(K) \to \mathrm{Ban}_G(K)$.

Proposition 4.20 *The functor*

$$\mathrm{Ban}_G(K) \to \mathrm{Ban}_G(K)$$

$$V \mapsto V'$$

is exact.

Proof We start with an exact sequence of K-Banach space representations of G

$$0 \longrightarrow U \overset{\varphi}{\longrightarrow} V \overset{\psi}{\longrightarrow} W \longrightarrow 0.$$

Using standard arguments (such as those in the proof of Theorem 28 in Section 10.5 of [28]), we can show that

$$0 \longrightarrow W' \overset{\psi'}{\longrightarrow} V' \overset{\varphi'}{\longrightarrow} U'$$

is an exact sequence of K-Banach space representations of G. To prove that φ' is surjective, we identify U with its image in V. Then $\varphi' : V' \to U'$ is the restriction $\lambda \mapsto \lambda|_U$. By Hahn-Banach theorem, any continuous linear functional on U extends to V (see Corollary 9.4 in [60]). It follows that the restriction $\lambda \mapsto \lambda|_U$ is surjective. Going back to original U, we see that $\varphi' : V' \to U'$ is surjective. $\qquad\square$

4.3 Schneider-Teitelbaum Duality

The duality theory of Schneider and Teitelbaum is built on Schikhof's duality between p-adic Banach spaces and compactoids [58].

4.3.1 Schikhof's Duality

Following Section 1 in [64], we will describe two functors $M \mapsto M^d$ and $V \mapsto V^d$. The first one is on the category

$$\mathrm{Mod}^{\mathrm{fl}}_{\mathrm{comp}}(o_K)$$

of all torsionfree and compact Hausdorff linear-topological o_K-modules, with morphisms being all continuous o_K-linear maps. For any module M in $\mathrm{Mod}^{\mathrm{fl}}_{\mathrm{comp}}(o_K)$, define

$$M^d = \mathrm{Hom}^c_{o_K}(M, K),$$

the space of all continuous o_K-linear maps $\ell : M \to K$. This is a K-Banach space with respect to the norm $\|\ell\| = \max_{m \in M} |\ell(m)|$. If M and N are modules in $\mathrm{Mod}^{\mathrm{fl}}_{\mathrm{comp}}(o_K)$ and $\varphi : M \to N$ is a continuous o_K-linear map, then $\varphi^d : N^d \to M^d$ is given by $\varphi^d(\ell) = \ell \circ \varphi$. Notice that for any $\ell \in N^d$,

$$\|\varphi^d(\ell)\| = \max_{m \in M} |\ell(\varphi(m))| \leq \max_{n \in N} |\ell(n)| = \|\ell\|.$$

It follows that φ^d is norm decreasing. Let

$$\text{Norm}(K)^{\leq 1}$$

denote the category of all complete normed spaces $(V, \| \ \|)$ such that $\|V\| \subset |K|$, with morphisms being all norm-decreasing K-linear maps. Then

$$M \mapsto M^d$$

is a functor from $\text{Mod}^{\text{fl}}_{\text{comp}}(o_K)$ to $\text{Norm}(K)^{\leq 1}$.

Remark 4.21 In [58], Schikhof works with normpolar normed spaces over non-archimedean fields. Since the valuation of K is discrete, a complete normed space $(V, \| \ \|)$ is normpolar if and only if $\|V\| \subset |K|$ (see Section 2.3 in [58]).

Lemma 4.22 *For $V \in \text{Norm}(K)^{\leq 1}$, let*

$$V^d = \{\lambda \in V' \mid \|\lambda\| \leq 1\}$$

be the unit ball of the dual space V' equipped with the weak topology. Then $V^d \in \text{Mod}^{\text{fl}}_{\text{comp}}(o_K)$.

Proof Let $V_0 = \{v \in V \mid \|v\| \leq 1\}$ be the unit ball in V. Then for every $\lambda \in V^d$ and $v \in V_0$, we have $|\lambda(v)| \leq 1$. Hence, we can define a map

$$\iota : V^d \to \prod_{v \in V_0} o_K$$

$$\lambda \mapsto (\lambda(v))_v.$$

It is clearly o_K-linear. For injectivity, take $\lambda_1, \lambda_2 \in V^d$ such that $\lambda_1 \neq \lambda_2$. Then, there exists $v \in V$ such that $\lambda_1(v) \neq \lambda_2(v)$. Let v_0 be a nonzero scalar multiple of v which belongs to V_0. Then $\lambda_1(v_0) \neq \lambda_2(v_0)$, and consequently $\iota(\lambda_1) \neq \iota(\lambda_2)$.

To prove that ι is continuous, take an open neighborhood U of zero in $\prod_{v \in V_0} o_K$. It contains an open neighborhood of the form $\prod_{v \in V_0} U_v$ where $U_v = o_K$ for all but finitely many $v_1, \ldots, v_s$, and for those v_i, $U_{v_i} = \mathfrak{p}_K^{n_i}$. Let $m = \max_i n_i$. Then

$$\iota(\mathcal{L}(\{v_1, \ldots, v_s\}, \mathfrak{p}_K^m)) \subset U,$$

proving that ι is continuous.

We claim that ι is closed. To prove the claim, take $a = (a_v)_{v \in V_0} \in \prod_{v \in V_0} o_K$ such that $a \notin \iota(V^d)$. Then there exists $S = \{v_0, v_1, \ldots, v_n\} \subset V_0$ such that

$$v_0 = b_1 v_1 + \cdots + b_n v_n \quad \text{but} \quad a_{v_0} \neq b_1 a_{v_1} + \cdots + b_n a_{v_n}.$$

Take $\epsilon > 0$ such that $|a_{v_0} - (b_1 a_{v_1} + \cdots + b_n a_{v_n})| > \epsilon$. For $i = 0, 1, \ldots, n$, let U_i be an open neighborhood of a_{v_i} such that for any $x \in U_i$ we have $|b_i(x - a_{v_i})| < \epsilon$ (with $b_0 = 1$). Let

$$\mathcal{U} = \prod_{v_i \in S} U_i \times \prod_{v \notin S} o_K.$$

This is an open neighborhood of $a = (a_v)_{v \in V_0}$ in $\prod_{v \in V_0} o_K$. We will show that $\mathcal{U}$ is disjoint from $\iota(V^d)$. Take $c = (c_v)_{v \in V_0} \in \iota(V^d)$. Then $c = \iota(\lambda)$ for some λ, so $c_v = \lambda(v)$, for all v. Suppose $c_{v_i} \in U_i$, for $i = 1, \ldots, n$. By linearity of λ, we have $c_{v_0} = b_1 c_{v_1} + \cdots + b_n c_{v_n}$. Then

$$c_{v_0} - a_{v_0} = b_1(c_{v_1} - a_{v_1}) + \cdots + b_n(c_{v_n} - a_{v_n}) + (b_1 a_{v_1} + \cdots + b_n a_{v_n}) - a_{v_0}.$$

Since

$$|b_1(c_{v_1} - a_{v_1}) + \cdots + b_n(c_{v_n} - a_{v_n})| \leq \max_{i \in \{1, \ldots, n\}} |b_i(c_{v_i} - a_{v_i})| < \epsilon$$

and $|a_{v_0} - (b_1 a_{v_1} + \cdots + b_n a_{v_n})| > \epsilon$, it follows from Lemma A.11 that $|c_{v_0} - a_{v_0}| > \epsilon$. Then $c_{v_0} \notin U_0$ and $c \notin \mathcal{U}$. This proves that ι is closed.

As a direct product of compact spaces, $\prod_{v \in V_0} o_K$ is compact. It follows that V^d is compact. □

Recall that for two categories C and D, an **equivalence of categories** consists of a functor $F : C \to D$ and a functor $G : D \to C$ such that GF is naturally isomorphic to I_C (the identity functor of C) and FG is naturally isomorphic to I_D. In this situation, F and G are said to be **quasi-inverses**. If the functors F and G are contravariant, we will use the term **anti-equivalence of categories**, also called a **duality of categories**. Here is another characterization of the equivalence of categories (see [45, IV.4, Theorem 1] or [55, Theorem 1.5.9]).

Lemma 4.23 *A functor $F : C \to D$ yields an equivalence of categories if and only if it is simultaneously:*

(i) *full, i.e. for any two objects A and B of C, the map $\mathrm{Hom}_C(A, B) \to \mathrm{Hom}_D(FA, FB)$ induced by F is surjective;*

(ii) *faithful, i.e. for any two objects A and B of C, the map $\mathrm{Hom}_C(A, B) \to \mathrm{Hom}_D(FA, FB)$ induced by F is injective; and*

(iii) *essentially surjective (dense), i.e. each object A in D is isomorphic to an object of the form FB, for B in C.*

Lemma 4.24 *If $V \in \mathrm{Norm}(K)^{\leq 1}$ then $V^{dd} \cong V$.*

Proof By definition, V^d is the unit ball of the dual space V' equipped with the weak topology and $V^{dd} = \mathrm{Hom}_{o_K}^c(V^d, K)$. We claim that

$$\mathrm{Hom}_{o_K}^c(V^d, K) = \mathrm{Hom}_K^c(V'_{bs}, K).$$

One inclusion is trivial: if $f \in \operatorname{Hom}_K^c(V'_{bs}, K)$, then clearly $f|_{V^d} \in \operatorname{Hom}_{o_K}^c(V^d, K)$. For the converse inclusion, take $f_0 \in \operatorname{Hom}_{o_K}^c(V^d, K)$. We can extend it linearly to a K-linear map $f \in \operatorname{Hom}_K(V'_{bs}, K)$. We have to prove that f is continuous. Let U be an open lattice in K. Then $\mathcal{L}_0 = f_0^{-1}(U)$ is an open lattice in V^d. Moreover, for every $n \geq 0$, the set $\mathcal{L}_n = f_0^{-1}(\varpi_K^n U)$ is an open lattice in V^d. Define

$$\mathcal{L} = \bigcup_{n \geq 0} \varpi_K^{-n} \mathcal{L}_n.$$

Take $\ell \in \mathcal{L}$. Then $\ell = \varpi_K^{-n} \ell_0$ for some n and some $\ell_0 \in \mathcal{L}_n$, and we have $f(\ell) = \varpi_K^{-n} f_0(\ell_0) \in U$. This proves that $f(\ell) \in U$, for all $\ell \in \mathcal{L}$.

Notice that $\mathcal{L} \cap \varpi_K^{-n} V^d = \varpi_K^{-n} \mathcal{L}_n$ because if $\ell \in \mathcal{L} \cap \varpi_K^{-n} V^d$, then $f(\ell) \in U$ implies $f_0(\varpi_K^n \ell) = \varpi_K^n f(\ell) \in \varpi_K^n U$ and $\ell \in \varpi_K^{-n} \mathcal{L}_n$. The other containment is trivial. Lemma 3.36 tells us that the lattice $\mathcal{L}$ is open in V'_{bs}, thus proving that f is continuous at zero. By linearity, it is continuous. Hence,

$$V^{dd} = \operatorname{Hom}_{o_K}^c(V^d, K) = \operatorname{Hom}_K^c(V'_{bs}, K) = (V'_{bs})'.$$

The norm on V^{dd} is given by $\|f_0\| = \max_{\ell \in V^d} |f_0(\ell)|$. By Lemma 3.12, this is the operator norm on $(V'_{bs})'$. It follows $V^{dd} = (V'_{bs})'_b$. This is isomorphic to V by Corollary 3.40. □

Theorem 4.25 *The functor*

$$\operatorname{Mod}_{\mathrm{comp}}^{\mathrm{fl}}(o_K) \mapsto \operatorname{Norm}(K)^{\leq 1}$$

$$M \mapsto M^d$$

is an anti-equivalence of categories, with quasi-inverse $V \mapsto V^d$.

Proof This follows from the proof of Theorem 1.2 in [64]. The authors prove that the o_K-linear map

$$\iota_M : M \to (M^d)'_s$$

given by $m \mapsto [\lambda \mapsto \lambda(m)]$ has image M^{dd} and that $\iota_M : M \to M^{dd}$ is a topological isomorphism. On the other hand, $V^{dd} \cong V$ by Lemma 4.24. □

Let A be an additive category. Then all its hom-sets are abelian groups and composition of morphisms is bilinear. It follows that we can define the additive category $A_{\mathbb{Q}}$ with the same objects as A and such that

$$\operatorname{Hom}_{A_{\mathbb{Q}}}(X, Y) = \operatorname{Hom}_A(X, Y) \otimes_{\mathbb{Z}} \mathbb{Q}$$

for any two objects X, Y in A. The composition of morphisms in $A_{\mathbb{Q}}$ is the $\mathbb{Q}$-linear extension of the composition in A.

Lemma 4.26 *The categories* $(\mathrm{Norm}(K)^{\leq 1})_{\mathbb{Q}}$ *and* $\mathrm{Ban}(K)$ *are equivalent.*

Proof Clearly, we have a forgetful functor $\mathrm{Norm}(K)^{\leq 1} \xrightarrow{\text{forget}} \mathrm{Ban}(K)$. Also, if f is a morphism in $\mathrm{Norm}(K)^{\leq 1}$ and $a \in \mathbb{Q}$, then af is a morphism in $\mathrm{Ban}(K)$. This gives us the functor

$$F : (\mathrm{Norm}(K)^{\leq 1})_{\mathbb{Q}} \longrightarrow \mathrm{Ban}(K).$$

Set $A = (\mathrm{Norm}(K)^{\leq 1})_{\mathbb{Q}}$ and $B = \mathrm{Ban}(K)$. We will show that $F : A \to B$ is fully faithful and essentially surjective, and hence an equivalence of categories (see Lemma 4.23).

From Lemma 4.1, we know that for every K-Banach space V there exists a norm $\| \ \|$ on V inducing the Banach space topology such that $\|V\| \subset |K|$. We abuse the notation and write $V = FV$. This shows that F is essentially surjective.

Take $V, W \in \mathrm{Obj}(A)$ and consider the map

$$\mathrm{Hom}_A(V, W) \longrightarrow \mathrm{Hom}_B(FV, FW) \tag{4.1}$$

induced by F. Given $f \in \mathrm{Hom}_A(V, W)$, we have $Ff = f$, considered as a map in $\mathrm{Hom}_B(FV, FW)$. Therefore, (4.1) is injective and F is faithful.

To show that F is full, take $f \in \mathrm{Hom}_B(FV, FW) = \mathcal{L}(V, W)$. If $\|f\| > 1$, there exists $k \in \mathbb{N}$ such that $\|p^k f\| \leq 1$. If $\|f\| \leq 1$, take $k = 0$. Set $f_0 = p^k f$. Then f_0 is a morphism in the category $\mathrm{Norm}(K)^{\leq 1}$ and $f = p^{-k} f_0 \in \mathrm{Hom}_A(V, W)$. This proves that (4.1) is surjective, so F is full, completing the proof. $\square$

Corollary 4.27 *The functor*

$$\mathrm{Mod}^{\mathrm{fl}}_{\mathrm{comp}}(o_K)_{\mathbb{Q}} \mapsto \mathrm{Ban}(K)$$

$$M \mapsto M^d$$

is an anti-equivalence of categories.

When applying Corollary 4.27 on Banach space representations, we will also consider topologies on hom-sets. If M and N are two modules in $\mathrm{Mod}^{\mathrm{fl}}_{\mathrm{comp}}(o_K)$, then the natural topology to consider on $\mathrm{Hom}^c_{o_K}(M, N)$ is the compact-open topology.

Suppose V and W are K-Banach spaces. As in the proof of Lemma 4.26, we can consider V and W as objects in $(\mathrm{Norm}(K)^{\leq 1})_{\mathbb{Q}}$ and define V^d and W^d. By the above equivalence of categories we have a natural linear isomorphism

$$\mathcal{L}(V, W) \xrightarrow{\sim} \mathrm{Hom}^c_{o_K}(W^d, V^d) \otimes \mathbb{Q}.$$

Lemma 4.28 *The bounded weak topology on $\mathcal{L}(V, W)$ induces the compact-open topology on $\mathrm{Hom}^c_{o_K}(W^d, V^d)$.*

Proof This is Proposition 8.1 in [62]. □

4.3.2 Duality for Banach Space Representations: Iwasawa Modules

In this section, G_0 is a profinite group. We want to apply Schikhof duality on Banach space representations of G_0. We start with Lemma 2.1 and Corollary 2.2 from [64]. The proofs are omitted and can be found in [64].

Lemma 4.29 *Let M be a complete Hausdorff linear-topological o_K-module. Then the restriction map $f \mapsto f|_{G_0}$ defines a bijection*

$$\mathrm{Hom}^c_{o_K}(o_K[[G_0]], M) \xrightarrow{\sim} C(G_o, M).$$

A locally convex vector space V is called quasi-complete if every bounded closed subset of V is complete (see [60, §7]).

Corollary 4.30 *For any quasi-complete Hausdorff locally convex K-vector space V we have the K-linear isomorphism*

$$\mathcal{L}(K[[G_0]], V) \xrightarrow{\sim} C(G_o, V)$$

$$f \mapsto f|_{G_0}.$$

Example 4.31 (Dual Pairing) Recall that $o_K[[G_0]] \cong D^c(G_0, o_K)$ and $K[[G_0]] \cong D^c(G_0, K)$, with both isomorphisms obtained from $g \mapsto \delta_g$, for $g \in G_0$ (Theorems 3.32 and 3.44). Then we can identify $K[[G_0]]$ and $D^c(G_0, K)$. The canonical pairing $\langle , \rangle : D^c(G_0, K) \times C(G_0, K) \to K$ induces

$$\langle , \rangle : K[[G_0]] \times C(G_0, K) \to K.$$

In Sect. 3.2.3, we explained how to compute $\langle \mu, f \rangle$, for $\mu \in o_K[[G_0]]$ and $C(G_0, o_K)$. This extends linearly to get $\langle \mu, f \rangle$ for $\mu \in K[[G_0]]$ and $f \in C(G_0, K)$.

Now, we look at the dual pairing from the point of view of Corollary 4.30. Let $f \in C(G_0, K)$. Set $F = \langle , f \rangle$. For $g \in G_0$, we have

$$F(g) = \langle g, f \rangle = f(g),$$

and therefore $F|_{G_0} = f$. Then $F = \langle , f \rangle$ is the unique continuous linear map $F : K[[G_0]] \to K$ such that $F|_{G_0} = f$.

$K[[G_0]]$-module structure on V'

Let V be a K-Banach space representation of G_0. By Definition 4.13, we have a continuous homomorphism

$$\pi : G_0 \to \mathcal{L}_s(V, V),$$

so $\pi \in C(G_0, \mathcal{L}_s(V, V))$. From [60, Corollary 7.14], we know that $\mathcal{L}_s(V, V)$ is quasi-complete and Hausdorff. Then, by Corollary 4.30, π extends uniquely to a continuous K-linear map

$$\Pi : K[[G_0]] \to \mathcal{L}_s(V, V), \quad \Pi|_{G_0} = \pi.$$

On the other hand, we can extend π K-linearly to $K[G_0]$. We denote the resulting map again by π. Then $\pi : K[G_0] \to \mathcal{L}_s(V, V)$ is a K-algebra homomorphism. By K-linearity, we have $\Pi|_{K[G_0]} = \pi$. Since $K[G_0]$ is dense in $K[[G_0]]$, it follows that Π is also a K-algebra homomorphism. Hence, we have a continuous algebra homomorphism

$$K[[G_0]] \to \mathcal{L}_s(V, V).$$

Since $o_K[[G_0]]$ is compact, the corresponding map

$$K[[G_0]] \to \mathcal{L}_{bs}(V, V) \tag{4.2}$$

is also a continuous algebra homomorphism.

Next, we need a G_0-invariant norm inducing the topology on V. Since G_0 is compact, such a norm always exists. We remark that for non-compact Lie groups the lemma below is no longer true.

Lemma 4.32 *Let V be a K-Banach space representation of G_0. Then there is a G_0-invariant norm inducing the topology on V.*

Proof We follow [62, Remark 8.2]. As a profinite group, G_0 has a neighborhood basis of the identity consisting of compact open normal subgroups (see Proposition 2.32). Take a bounded open lattice $\mathcal{L}$ in V. Since the action $G_0 \times V \to V$ is continuous, there exist a compact open normal subgroup H of G_0 and an open lattice $\mathcal{L}_0 \subseteq \mathcal{L}$ such that $H \cdot \mathcal{L}_0 \subseteq \mathcal{L}$. Set

$$\mathcal{L}_1 = \bigcap_{h \in H} h\mathcal{L}.$$

For any $h \in H$ we have $h^{-1} \cdot \mathcal{L}_0 \subseteq \mathcal{L}$ and hence $\mathcal{L}_0 \subseteq h \cdot \mathcal{L}$. It follows $\mathcal{L}_0 \subseteq \mathcal{L}_1$, so $\mathcal{L}_1$ is also an open lattice in V. The lattice $\mathcal{L}_2 = \bigcap_{g \in G_0} g\mathcal{L}$ is G_0-invariant. It is open because

$$\mathcal{L}_2 = \bigcap_{g \in G_0} g\mathcal{L} = \bigcap_{g \in G_0/H} g\mathcal{L}_1,$$

so $\mathcal{L}_2$ is a finite intersection of open lattices $g\mathcal{L}_1$.

Having constructed an open G_0-invariant lattice, the corresponding norm is also G_0-invariant. More specifically, we define

$$\|v\|_2 = \inf_{v \in a\mathcal{L}_2} |a|.$$

Then $\|\ \|_2$ is a G_0-invariant norm on V which defines the topology [60, Proposition 4.4].   □

Lemma 4.33 *Let V be a K-Banach space representation of G_0. Let $\|\ \|$ be a G_0-invariant norm on V inducing the Banach space topology and let $M = V^d$ be the unit ball in V' equipped with the weak topology.*

(i) The G_0-action on M induces a continuous map $o_K[[G_0]] \times M \to M$.
(ii) The G_0-action on V'_{bs} induces a separately continuous map $K[[G_0]] \times V'_{bs} \to V'_{bs}$.

Proof As explained above, the map $K[[G_0]] \to \mathcal{L}_{bs}(V, V)$ from Eq. (4.2) is a continuous homomorphism. Since $\|\ \|$ is G_0-invariant, Lemma 4.28 gives us a continuous homomorphism $o_K[[G_0]] \to \mathrm{Hom}^c_{o_K}(M, M)$, where $\mathrm{Hom}^c_{o_K}(M, M)$ carries the compact-open topology. Then Theorem 3 in [10, Ch. X, §3.4] implies that $o_K[[G_0]] \times M \to M$ is continuous. This proves (i).

Assertion (ii) follows from (i). The proof is left as an exercise.   □

Let M be an $o_K[[G_0]]$-module. Then M also carries the underlying o_K-module structure coming from the embedding $o_K \subset o_K[[G_0]]$.

Definition 4.34 Let G_0 be a profinite group.

(i) An $o_K[[G_0]]$-**Iwasawa module** is a topological $o_K[[G_0]]$-module M such that the underlying o_K-module belongs to $\mathrm{Mod}^{\mathrm{fl}}_{\mathrm{comp}}(o_K)$.
(ii) A $K[[G_0]]$-Iwasawa module is a locally convex K-vector space U equipped with a separately continuous action $K[[G_0]] \times U \to U$.

If M is an $o_K[[G_0]]$-Iwasawa module then, by definition of a topological $o_K[[G_0]]$-module, the action $o_K[[G_0]] \times M \to M$ is continuous.

We denote by

$$\mathrm{Mod}^{\mathrm{fl}}_{\mathrm{comp}}(o_K[[G_0]])$$

the category of all $o_K[[G_0]]$-Iwasawa modules with morphisms being all continuous $o_K[[G_0]]$-module homomorphisms. Then Theorem 2.3 in [64] states the following.

Theorem 4.35 *The functor*

$$\text{Mod}^{\text{fl}}_{\text{comp}}(o_K[[G_0]])_\mathbb{Q} \mapsto \text{Ban}_{G_0}(K)$$

$$M \mapsto M^d$$

is an anti-equivalence of categories.

Remark 4.36 In [64], Schneider and Teitelbaum mention the following two pathologies of the category $\text{Ban}_{G_0}(K)$:

(i) There exist non-isomorphic topologically irreducible K-Banach space representations V and W of G_0 such that $\text{Hom}^c_{G_0}(V, W) \neq 0$.
(ii) The group $\mathbb{Z}_p$, which is a compact abelian group, still has a huge number of infinite dimensional topologically irreducible Banach space representations (see Diarra [27]).

To avoid such pathologies, they introduce an additional finiteness condition called admissibility. In the next section, we will define the category $\text{Ban}^{\text{adm}}_{G_0}(K)$ of admissible K-Banach space representations of G_0. This category avoids both of the above listed pathologies. For the first one, this follows from Corollary 4.47. For the second one, we refer to the discussion on page 375 of [64].

4.4 Admissible Banach Space Representations

In this section, G_0 is a compact Lie group. Admissible Banach space representations of G_0, as in Definition 4.41, form an important subcategory of $\text{Ban}_{G_0}(K)$. This subcategory is algebraic in nature and its definition is based on the following fundamental property of Iwasawa algebras.

Theorem 4.37 *Let G_0 be a compact Lie group. Then*

(i) The ring $o_K[[G_0]]$ is left and right noetherian.
(ii) The ring $K[[G_0]]$ is left and right noetherian.

Proof Assertion (i) was proved by Lazard in [44, V.2.2.4] (also, cf. [63, Theorem 33.4]). Assertion (ii) follows from (i). □

Since $K[[G_0]]$ is noetherian, the category

$$\text{Mod}_{\text{fg}}(K[[G_0]]).$$

of all finitely generated $K[[G_0]]$-Iwasawa modules has nice algebraic properties.

Proposition 4.38 *Let G_0 be a compact Lie group. Then*

(i) *Any finitely generated $o_K[[G_0]]$-module M carries a unique Hausdorff topology—its canonical topology—such that the action $o_K[[G_0]] \times M \to M$ is continuous.*

(ii) *Any submodule of a finitely generated $o_K[[G_0]]$-module is closed in the canonical topology.*

(iii) *Any $o_K[[G_0]]$-linear map between two finitely generated $o_K[[G_0]]$-modules is continuous in the canonical topologies.*

Proof Follows from the fact that $o_K[[G_0]]$ is compact and noetherian.   □

Exercise 4.39 Prove Proposition 4.38.

Lemma 4.40 *Let G_0 be a compact Lie group, and let M be a $K[[G_0]]$-module. Then for any compact open subgroup H of G_0, M is finitely generated as a $K[[H]]$-module if and only if it is finitely generated as a $K[[G_0]]$-module.*

Proof From Proposition 2.54 (ii), we know that $K[[H]]$ is a closed subalgebra of $K[[G_0]]$. If M is finitely generated as a $K[[H]]$-module, then it follows trivially that M is finitely generated as a $K[[G_0]]$-module.

For the converse statement, let $\{g_1, \ldots, g_s\}$ be a set of coset representatives of $H \backslash G_0$. From Proposition 2.54 (iv), tensoring with K, we obtain

$$K[[G_0]] = K[[H]]g_1 \oplus \cdots \oplus K[[H]]g_s.$$

If M is finitely generated as a $K[[G_0]]$-module, with generators $m_1, \ldots, m_r$, then $\{g_i m_j \mid i = 1, \ldots, s, \ j = 1, \ldots, r\}$ generate M as a $K[[H]]$-module.   □

Definition 4.41 Let G be a p-adic Lie group. A K-Banach space representation V of G is called **admissible** if the dual space V' is finitely generated as a $K[[H]]$-module for some compact open subgroup H of G.

Suppose that V is an admissible K-Banach space representation of G. It follows from Lemma 4.40 that V' is finitely generated as a $K[[H]]$-module for **any** compact open subgroup H of G. We remark that Definition 4.41 is not the original definition of admissibility, but an equivalent property given in Lemma 3.4 of [64].

We denote by

$$\mathrm{Ban}_G^{\mathrm{adm}}(K)$$

the full subcategory in $\mathrm{Ban}_G(K)$ of all admissible representations.

Let G_0 be a compact p-adic Lie group and V an admissible K-Banach space representation of G_0. By definition of admissibility, V corresponds under duality to the finitely generated $K[[G_0]]$-module V'.

Lemma 4.42 *Suppose U and W are finitely generated $K[[G_0]]$-Iwasawa modules. If $f \in \mathrm{Hom}_{K[[G_0]]}(U, V)$, then f is continuous.*

Proof Exercise.   □

Theorem 4.43 *Let G_0 be a compact p-adic Lie group. The functor*

$$\mathrm{Ban}_{G_0}^{\mathrm{adm}}(K) \mapsto \mathrm{Mod}_{\mathrm{fg}}(K[[G_0]])$$

$$V \mapsto V'$$

is an anti-equivalence of categories.

Proof This follows from Theorem 3.5 in [64].   □

Corollary 4.44 *The categories $\mathrm{Mod}_{\mathrm{fg}}(K[[G_0]])$ and $\mathrm{Ban}_{G_0}^{\mathrm{adm}}(K)$ are abelian.*

Proof Since $K[[G_0]]$ is noetherian, it follows that $\mathrm{Mod}_{\mathrm{fg}}(K[[G_0]])$ is abelian. Theorem 4.43 implies that $\mathrm{Ban}_{G_0}^{\mathrm{adm}}(K)$ is also abelian.   □

Lemma 4.45 *Let V be an admissible K-Banach space representation of G_0. Then V is topologically irreducible if and only if V' is a simple $K[[G_0]]$-module.*

Proof If $U \neq 0$ is a proper closed G_0-invariant subspace, then we have the exact sequence of Banach space representations $0 \to U \to V \to V/U \to 0$ and the corresponding dual sequence of $K[[G_0]]$-modules

$$0 \to (V/U)' \to V' \to U' \to 0$$

(see Proposition 4.20). If V' is a simple $K[[G_0]]$-module, then V must be topologically irreducible representation of G_0.

For the converse, suppose that V' contains a proper $K[[G_0]]$-submodule $S \neq 0$. Let $\| \ \|$ be a G_0-invariant norm on V inducing the Banach space topology such that $\|V\| \subset |K|$. Let M be the unit ball in V' equipped with the weak topology. Then $M = V^d$, and hence M is compact by Lemma 4.22. Let $\lambda_1, \ldots, \lambda_s$ be a set of generators of V', so $V' = K[[G_0]]\lambda_1 + \cdots + K[[G_0]]\lambda_s$. By scaling, using $\|V\| \subset |K|$, we may assume $\|\lambda_i\| = 1$, for all $i = 1, \ldots, s$. Define

$$N = S \cap (o_K[[G_0]]\lambda_1 + \cdots + o_K[[G_0]]\lambda_s).$$

This is a proper $o_K[[G_0]]$-submodule of M. It is closed in M by Proposition 4.38, and $M/N \neq 0$. It follows from Proposition 1.3 in [64] that the kernel of the dual map $M^d \to N^d$ is a nonzero proper closed G_0-invariant subspace of $M^d = V^{dd}$. Since $V^{dd} \cong V$ by Lemma 4.24, it follows that V contains a proper closed G_0-invariant subspace.   □

From Lemma 4.45, we obtain the following simple and operative formulation of the Schneider-Teitelbaum duality:

Corollary 4.46 (Schneider-Teitelbaum Duality) *Let G_0 be a compact p-adic Lie group. The functor*

$$V \mapsto V'$$

induces a bijection between the set of isomorphism classes of topologically irreducible admissible K-Banach space representations of G_0 and the set of isomorphism classes of simple $K[[G_0]]$-modules.

The following corollary addresses Remark 4.36 (i).

Corollary 4.47 *Any nonzero G_0-equivariant map between two topologically irreducible admissible K-Banach space representations of G_0 is an isomorphism.*

The duality given in Corollary 4.46 is powerful. It enables us to work with Banach space representations using (mostly) purely algebraic methods. This approach will be taken in Part II, for studying principal series representations. We will also need a formulation for non-compact groups, and it is given in Sect. 4.4.2.

4.4.1   Locally Analytic Vectors: Representations in Characteristic p

In this section, we mention briefly two types of representations related to Banach space representations. Our discussion is informal.

Locally Analytic Vectors

Let V be a K-Banach space. If G is a Lie group over $L \subseteq K$, we denote by $C^{an}(G, V)$ the space of locally L-analytic functions $f : G \to V$.

Let (π, V) be a K-Banach space representation of G. A vector $v \in V$ is called *locally L-analytic* if the map $g \mapsto \pi(g)v$ lies in $C^{an}(G, V)$. A vector $v \in V$ is called *smooth* if there exists an open subgroup H of G such that $\pi(h)v = v$ for all $h \in H$ (see Definition 6.12). Denote by V^{sm} the subspace of smooth vectors of V and by V^{an} the subspace of locally L-analytic vectors of V. Then we have the following sequence of G-invariant subspaces

$$V^{sm} \subseteq V^{an} \subseteq V,$$

where both V^{sm} and V^{an} may be zero. In the case when $L = \mathbb{Q}_p$ and V is admissible, the space of locally analytic vectors V^{an} is dense in V [66, Theorem 7.1]. The proof is based on algebraic properties of distribution algebras. Namely, for a compact p-adic Lie group G_0, the strong dual of $C^{an}(G_0, K)$ is denoted by

$$D(G_0, K) = C^{an}(G_0, K)'_b$$

and it is called the algebra of locally $\mathbb{Q}_p$-analytic distributions on G_0. Then $D^c(G_0, K)$ is dense in $D(G_0, K)$ [65, Lemma 3.1] and the natural ring homomorphism $D^c(G_0, K) \to D(G_0, K)$ is faithfully flat [66, Theorem 5.2].

Unitary Representations and Reduction Modulo $\mathfrak{p}_K$

Let G be a p-adic Lie group and let (π, V) be a Banach space representation of G. We say that π is **unitary** if the topology on V is defined by a G-invariant norm $\| \ \|$. Then

$$\|\pi(g)v\| = \|v\| \quad \text{for all } g \in G, v \in V. \tag{4.3}$$

Remark 4.48 It would be more appropriate to call such representations *isometric*. The term *unitary representation* usually refers to a representation of G on a complex vector space V such that there is a Hermitian inner product on V that is G-invariant (see [70, §1.3] or [53, IV.2.2.]). However, since the term *unitary* is used for norm-preserving representations in many important works on p-Banach space representations, we will use the latter terminology.

Suppose (π, V) is an admissible unitary K-Banach space representation of G. Let $\kappa = o_K/\mathfrak{p}_K$ be the residue field of K, as defined in Appendix A.2.2. Then κ is a finite extension of $\mathbb{F}_p$. Consider the lattice

$$V_0 = \{v \in V \mid \|v\| \le 1\}.$$

By Eq. (4.3), V_0 is G-invariant, and so is $\varpi_K V_0$. Then $\overline{V} = V_0/\varpi_K V_0$ is a smooth κ-linear representation of G.

4.4.2 Duality for p-adic Lie Groups

Let G be a p-adic Lie group, and G_0 a compact open subgroup of G. Suppose that V is an admissible K-Banach space representation of G. Then V' carries two actions: it is a G-module and also a finitely generated $K[[G_0]]$-module. These two actions coincide on G_0, leading to the following definition (similar to *augmented representations* defined by Emerton in [31]).

Definition 4.49 A $(K[[G_0]], G)$**-module** is a $K[[G_0]]$-Iwasawa module U equipped with a G-action $G \times U \to U$ such that the two actions coincide on G_0.

Recall that by definition of a $K[[G_0]]$-Iwasawa module, U is a locally convex K-vector space and the action $K[[G_0]] \times U \to U$ is separately continuous.

Theorem 4.50 (Schneider-Teitelbaum Duality II) *Let G be a p-adic Lie group, and G_0 a compact open subgroup of G. Let V be an admissible Banach space representation of G. The map*

$$U \mapsto U'$$

defines a bijection between the set of G-invariant closed vector subspaces of V and the set of G-invariant $K[[G_0]]$-quotient modules of V'.

Proof If $U \neq 0$ is a proper closed G-invariant subspace, then we have the exact sequence of Banach space representations $0 \to U \to V \to V/U \to 0$. From Proposition 4.20, we have the corresponding dual sequence

$$0 \to (V/U)' \to V' \to U' \to 0$$

which is $K[[G_0]]$-linear and G-equivariant. Then U' is a G-invariant $K[[G_0]]$-quotient module of V'. From the proof of Lemma 4.45, $U \mapsto U'$ is a bijection between the set of G_0-invariant closed vector subspaces of V and the set of $K[[G_0]]$-quotient modules of V'. Then we have the corresponding bijection of G-invariant objects on both sides, proving the theorem.   □

Remark 4.51 The finiteness condition in the definition of an admissible Banach space representation may seem very restrictive, as pointed to us by David Vogan. There are two aspects to consider:

(i) the size of an admissible K-Banach space representation, and
(ii) the size of the category $\mathrm{Ban}_G^{\mathrm{adm}}(K)$,

where we use the term *size* loosely.

In Chap. 7, we will define the continuous principal series $V = \mathrm{Ind}_P^G(\chi)$, where χ is a continuous character $\chi : T \to K^\times$. Then $V^{\mathrm{sm}} \neq 0$ if χ is smooth and $V^{\mathrm{an}} \neq 0$ if χ is locally L-analytic. For any χ, V is admissible (Corollary 7.13).

Suppose that χ is smooth. Then V^{sm}, considered as a G_0-representation, contains countably many finite-dimensional irreducible subrepresentations (see Sect. 8.3). Since they are finite-dimensional, they are also topologically irreducible. Then V contains countably many finite-dimensional topologically irreducible G_0-subrepresentations, and they comprise only a tiny part of V. Still, V' is a finitely generated $K[[G_0]]$-module, and V is admissible.

Continuous principal series are just a special type of admissible Banach space representations. The size and diversity of the category $\mathrm{Ban}_G^{\mathrm{adm}}(K)$ may be illustrated by the p-adic Langlands correspondence for $GL_2(\mathbb{Q}_p)$. To give a bigger picture, we will briefly abandon our good mathematical housekeeping of properly defining everything we talk about. The concepts discussed below are not used in the rest of the book.

Let (V, π) be an irreducible admissible K-Banach space representation of $GL_2(\mathbb{Q}_p)$. We say that V is *ordinary* if it is a subquotient of a continuous principal series induced from a unitary character.

Let $\psi : \mathcal{G}_{\mathbb{Q}_p} = \mathrm{Gal}(\overline{\mathbb{Q}}_p/\mathbb{Q}_p) \to GL_2(K)$ be an absolutely irreducible Galois representation. We denote by $\Pi(\psi)$ the unitary Banach space representation of $GL_2(\mathbb{Q}_p)$ attached to ψ by p-adic Langlands correspondence (Colmez [20], 0.17]). The functor

$$\psi \mapsto \Pi(\psi)$$

induces a bijection between the set of equivalence classes of absolutely irreducible continuous two-dimensional K-representations of $\mathcal{G}_{\mathbb{Q}_p}$ and absolutely irreducible non-ordinary admissible unitary K-Banach space representations of $GL_2(\mathbb{Q}_p)$ [22, 1.1]. Moreover, the p-adic Langlands correspondence encapsulates the classical Langlands correspondence; any irreducible admissible smooth representation σ of $GL_2(\mathbb{Q}_p)$ different from a character is a subrepresentation of $\Pi(\psi)$, for some ψ. Such ψ is necessarily de Rham, and from its Fontain-Deligne module we can obtain the L-parameter attached to σ by the classical Langlands correspondence.

More generally, $\Pi(\psi)$ may contain locally algebraic vectors. The space of locally algebraic vectors of $\Pi(\psi)$ is non-zero if and only if ψ is de Rham with distinct Hodge-Tate weights [20, Theorem 0.20].

As noted by Taylor in [75], "most" p-adic representations of $\mathcal{G}_{\mathbb{Q}_p}$ are not de Rham. So, "most" admissible unitary K-Banach space representations of $GL_2(\mathbb{Q}_p)$ do not contain smooth vectors or locally algebraic vectors. In summary, $\mathrm{Ban}_G^{\mathrm{adm}}(K)$ contains the subcategory of unitary representations V with $V^{\mathrm{sm}} \neq 0$, comparable in size with the category of all smooth representations of G, then the bigger subcategory of unitary representations V with nonzero locally algebraic vectors, and then the much bigger subcategory of unitary representations.

Part II
Principal Series Representations of Reductive Groups

Notation in Part II

In Part II, $\mathbb{Q}_p \subseteq L \subseteq K$ is a sequence of finite extensions. More general fields appear in Chap. 5 and Appendix: k is an algebraically closed field and F is a subfield of k.

We denote by $\mathbf{G}$ a split connected reductive $\mathbb{Z}$-group. Let $G = \mathbf{G}(L)$ be the group of L-points of $\mathbf{G}$.

We study principal series representations of G on K-Banach spaces. Our methods rely on the structure theory of G. Chapter 5 gives an overview of the structure theory of reductive groups. Below, we summarize the notation for some of the standard objects associated to G. These objects will be introduced in a systematic way in Chap. 5.

We take $\mathbf{P}$ a Borel subgroup of $\mathbf{G}$ and $\mathbf{T} \subset \mathbf{P}$ a maximal torus, and denote the unipotent radical of $\mathbf{P}$ by $\mathbf{U}$. The unipotent radical of the opposite parabolic is denoted by $\mathbf{U}^-$. We write Φ for the roots of $\mathbf{T}$ in $\mathbf{G}$ and Φ^+ (respectively, Φ^-) for the set of positive (respectively, negative) roots determined by the choice of $\mathbf{P}$. For each $\alpha \in \Phi$, we have the root subgroup $\mathbf{U}_\alpha$ and an isomorphism $x_\alpha : \mathbf{G}_a \to \mathbf{U}_\alpha$ from the additive group $\mathbf{G}_a$ to $\mathbf{U}_\alpha$.

We denote by o_L the ring of integers of L and by $\mathfrak{p}_L$ its unique maximal ideal. Let q_L be the cardinality of the residue field of L.

For each algebraic subgroup $\mathbf{H}$ of $\mathbf{G}$ we let $H = \mathbf{H}(L)$ and $H_0 = \mathbf{H}(o_L)$. We write pr_n for the canonical projection $o_L \to o_L/\mathfrak{p}_L^n$ and also for the induced map $H_0 \to \mathbf{H}(o_L/\mathfrak{p}_L^n)$ for any $\mathbf{H}$. The kernel of pr_n in H_0 is denoted H_n. Finally, we set $\bar{H} = H(o_L/\mathfrak{p}_L)$. Then $B = \mathrm{pr}_1^{-1}(\bar{P})$ is the standard Iwahori subgroup of G.

We denote by $W = W(\mathbf{G}, \mathbf{T})$ the Weyl group of $\mathbf{G}$ relative to $\mathbf{T}$. For each $w \in W$ we select a representative $\dot{w} \in \mathbf{G}(\mathbb{Z})$.

Chapter 5
Reductive Groups

Throughout Part II, $\mathbb{Q}_p \subseteq L \subseteq K$ is a sequence of finite extensions and $G = \mathbf{G}(L)$ is the group of L-points of a split connected reductive $\mathbb{Z}$-group $\mathbf{G}$.

In this chapter, we give an overview of the structure theory of split reductive $\mathbb{Z}$-groups. We avoid the language of algebraic geometry and base our presentation on the down-to-earth definition of an affine variety given in Appendix B. The purpose of this chapter is to help a learner navigate through the literature and to explain different objects associated to G, such as roots, unipotent subgroups, and Iwahori subgroups. We also review important structural results, such as Bruhat decomposition, Iwasawa decomposition, and Iwahori factorization. We do not try, however, to present all important results—we simply prepare for what will be needed in Chaps. 7 and 8.

For a systematically developed theory of reductive groups, we refer to the books by Borel [8], Humphreys [38], or Springer [72], all entitled *Linear Algebraic Groups*. We follow them in the first three sections. Sections 5.1 and 5.2 are over an algebraically closed field k, while in Sect. 5.3 we talk about F-groups and F-points, where F is a subfield of k.

In Sect. 5.4, we introduce $\mathbb{Z}$-groups, and in particular split reductive $\mathbb{Z}$-groups. Our approach rely on intuitive explanations, and on proper definitions given in the literature (Jantzen [40]). In Sect. 5.5, we describe the structure of $\mathbf{G}(L)$. It is richer than the structure over an arbitrary field F because, in addition to the structure of reductive groups, it also relies on the properties of nonarchimedean fields.

In this chapter, k is an algebraically closed field and F is a subfield of k.

5.1 Linear Algebraic Groups

We refer to Appendix B for the definition and basic properties of affine varieties. We just mention that an affine variety $V \subseteq \mathbf{A}^n = k^n$ is the set of common zeros of a

D. Ban, *p-adic Banach Space Representations*, Lecture Notes
in Mathematics 2325, https://doi.org/10.1007/978-3-031-22684-7_5

finite collection of polynomials in $k[x_1, \ldots, x_n]$. It carries the Zariski topology (see Definition B.6). Also, if $V \subset \mathbf{A}^n$ and $W \subset \mathbf{A}^m$ are two affine varieties, a morphism of affine varieties $\varphi : V \to W$ is defined by polynomial functions (see page 203).

Definition 5.1 A **linear algebraic group** is an affine variety $\mathbf{G}$ which is also a group such that the maps

$$\mu : \mathbf{G} \times \mathbf{G} \to \mathbf{G}, \quad \mu(a, b) = ab,$$
$$i : \mathbf{G} \to \mathbf{G}, \qquad i(a) = a^{-1}.$$

are morphisms of affine varieties.

A **morphism of linear algebraic groups** is a group homomorphism $\varphi : \mathbf{G} \to \mathbf{G}'$ which is also a morphism of affine varieties. Linear algebraic groups $\mathbf{G}$ and $\mathbf{G}'$ are **isomorphic** if there exists an isomorphism of varieties $\varphi : \mathbf{G} \to \mathbf{G}'$ which is also an isomorphism of groups.

5.1.1 Basic Properties of Linear Algebraic Groups

We start with basic examples of linear algebraic groups.

1. The **additive group** $\mathbf{G}_a$ is the affine line $\mathbf{A}^1 = k$ with addition as group operation.
2. The **multiplicative group** $\mathbf{G}_m$ is $k^\times$ with multiplication as group operation. To show that taking inverses is a polynomial function, we embed $\iota : \mathbf{G}_m \hookrightarrow \mathbf{A}^2$ as follows: for $x \in \mathbf{G}_m$, set $\iota(x) = (x, x^{-1})$. Let $k[x, y]$ be the affine algebra of $\mathbf{A}^2$. Then $\mathbf{G}_m$ is described as the set of zeros of the polynomial $xy - 1$ in $k[x, y]$. The map $x \mapsto x^{-1}$ on $\mathbf{G}_m$ is now just the restriction to $\mathbf{G}_m$ of the projection map $(x, y) \mapsto y$.
3. The **general linear group** $\mathbf{GL}_n$ is the group of all $n \times n$ matrices with coefficients in k and determinant $\neq 0$. If $n = 1$, then $\mathbf{GL}_1 = \mathbf{G}_m$.

   To show that $\mathbf{GL}_n$ is a linear algebraic group, we embed $\iota : \mathbf{GL}_n \hookrightarrow \mathbf{A}^{n^2+1}$ as follows: if $g = (g_{ij}) \in \mathbf{GL}_n$, set

$$\iota(g) = (g_{11}, \ldots, g_{nn}, (\det g)^{-1}) \in \mathbf{A}^{n^2+1}.$$

Let $k[x_{11}, \ldots, x_{nn}, y]$ be the affine algebra of $\mathbf{A}^{n^2+1}$. Then we can describe $\mathbf{GL}_n$ as the set of zeros of the polynomial

$$y(\det(x_{ij})) - 1.$$

It follows that $\mathbf{GL}_n$ is a closed subset of $\mathbf{A}^{n^2+1}$. The multiplication in $\mathbf{GL}_n$ is the product of matrices. It is given by polynomial functions, so it is a morphism of varieties. Also, the inverse on $\mathbf{GL}_n$ can be computed as $(x_{ij})^{-1} = y \, \mathrm{adj}(x_{ij})$, so it is a polynomial function. It follows that $\mathbf{GL}_n$ is a linear algebraic group.

A linear algebraic group is sometimes defined as a Zariski closed subgroup of $\mathbf{GL}_n$ for some n. This definition is equivalent to Definition 5.1, as we can see from the following theorem (Springer [72, Theorem 2.3.7]):

Theorem 5.2 *Let* $\mathbf{G}$ *be a linear algebraic group. Then there is an isomorphism of* $\mathbf{G}$ *onto a closed subgroup of some* $\mathbf{GL}_n$.

An element $g \in \mathbf{GL}_n$ is called **semisimple** if it is diagonalizable. It is called **unipotent** if $g - 1$ is nilpotent, that is, $(g - 1)^n = 0$ for some $n \in \mathbb{N}$.

Theorem 5.3 (Jordan Decomposition) *Let* $g \in \mathbf{GL}_n$. *There are unique elements* $g_s \in \mathbf{G}$ *and* $g_u \in \mathbf{G}$ *such that* g_s *is semisimple,* g_u *is unipotent, and* $g = g_s g_u = g_u g_s$.

More generally, Jordan decomposition holds in an arbitrary linear algebraic group (see Theorem 2.4.8 in [72]).

More Examples of Linear Algebraic Groups

1. We write $\mathrm{diag}(a_1, \ldots, a_n)$ for the diagonal $n \times n$ matrix with diagonal entries $a_1, \ldots, a_n$,

$$\mathrm{diag}(a_1, \ldots, a_n) = \begin{pmatrix} a_1 & & \\ & \ddots & \\ & & a_n \end{pmatrix}.$$

Let $\mathbf{D}_n$ be the subgroup of $\mathbf{GL}_n$ consisting of diagonal matrices,

$$\mathbf{D}_n = \{\mathrm{diag}(a_1, \ldots, a_n) \mid a_i \in k^\times\} \cong \mathbf{G}_m \times \cdots \times \mathbf{G}_m.$$

$\mathbf{D}_n$ is a closed subgroup of $\mathbf{GL}_n$. It is the set of zeros of the polynomials x_{ij}, $i \neq j$.

2. The group $\mathbf{T}_n$ of upper triangular invertible $n \times n$ matrices is a closed subgroup of $\mathbf{GL}_n$. It is the set of zeros of the polynomials x_{ij}, $i > j$.

3. Let $\mathbf{U}_n$ be the subgroup of $\mathbf{T}_n$ consisting of unipotent matrices. These are upper triangular matrices with all diagonal entries 1. Hence, if $g \in \mathbf{U}_n$, it is of the form

$$g = \begin{pmatrix} 1 & a_{12} & a_{13} & \ldots & a_{1n} \\ & 1 & a_{23} & \ldots & a_{2n} \\ & & \ddots & & \vdots \\ & & & & 1 \end{pmatrix}.$$

4. The **special linear group** $\mathbf{SL}_n$ consists of the matrices of determinant 1 in $\mathbf{GL}_n$.

5. The **symplectic group** $\mathbf{Sp}_{2n}$

$$\mathbf{Sp}_{2n} = \{g \in \mathbf{GL}_{2n} \mid {}^t g \begin{pmatrix} 0 & J \\ -J & 0 \end{pmatrix} g = \begin{pmatrix} 0 & J \\ -J & 0 \end{pmatrix}\},$$

where ${}^t g$ is the transpose of g and $J = \begin{pmatrix} & & 1 \\ & \cdot^{\cdot^{\cdot}} & \\ 1 & & \end{pmatrix}$.

6. The **special orthogonal group** $\mathbf{SO}_n$, $\mathrm{char}(k) \neq 2$, $\mathbf{SO}_n = \{g \in \mathbf{SL}_n \mid {}^t g J g = J\}$.

Unipotent Subgroups

The linear algebraic group $\mathbf{H}$ is called **unipotent** if all its elements are unipotent. The following is Proposition 2.4.12 from [72]:

Proposition 5.4 *Let* $\mathbf{H} \leq \mathbf{GL}_n$ *be a unipotent linear algebraic group. Then there exists* $g \in \mathbf{GL}_n$ *such that* $g\mathbf{H}g^{-1} \leq \mathbf{U}_n$.

Identity Component

Let $\mathbf{G}$ be an algebraic group. As an affine variety, $\mathbf{G}$ is disjoint union of irreducible varieties, $\mathbf{G} = V_0 \cup V_1 \cup \cdots \cup V_m$. Denote by $\mathbf{G}^0$ the unique irreducible component containing the identity element. We call $\mathbf{G}^0$ the **identity component** of $\mathbf{G}$. Then from Proposition 7.3 in [38], we have

Proposition 5.5 *Let* $\mathbf{G}$ *be an algebraic group. Then the identity component* $\mathbf{G}^0$ *is a normal subgroup of finite index in* $\mathbf{G}$. *The cosets of* $\mathbf{G}^0$ *are irreducible components of* $\mathbf{G}$.

An algebraic group $\mathbf{G}$ is called **connected** if $\mathbf{G} = \mathbf{G}^0$. The group $\mathbf{GL}_n$ is connected because it is a principal open set in an affine space. The classical groups $\mathbf{SL}_n$, $\mathbf{Sp}_{2n}$, and $\mathbf{SO}_n$ are also connected (see [38], page 53).

Example 5.6 The **even orthogonal group** $\mathbf{O}_{2n}$, for $\mathrm{char}(k) \neq 2$, is defined as $\mathbf{O}_{2n} = \{g \in \mathbf{GL}_n \mid {}^t g J g = J\}$. For any $g \in \mathbf{O}_{2n}$, we have ${}^t g J g = J$ and $\det J = \det({}^t g J g) = \det({}^t g) \det J \det g = \det J (\det g)^2$. It follows $(\det g)^2 = 1$ and $\det g = \pm 1$. The group $\mathbf{O}_{2n}$ has two irreducible components, $\mathbf{O}_{2n} = \mathbf{SO}_{2n} \cup s\mathbf{SO}_{2n}$, where s is an element of $\mathbf{O}_{2n}$ of determinant -1.

Tori

An algebraic group is called a **torus** if it is isomorphic to some $\mathbf{D}_n$.

Let $\mathbf{G}$ be a connected linear algebraic group. Then the maximal tori of $\mathbf{G}$ are all conjugate ([38], Section 21.3). The common dimension of the maximal tori of $\mathbf{G}$ is called the **rank** of $\mathbf{G}$.

Example 5.7 A maximal torus in $\mathbf{GL}_n$ is $\mathbf{D}_n$ and so rank$(\mathbf{GL}_n) = n$. Also, $\mathbf{SL}_n \cap \mathbf{D}_n$ is a maximal torus in $\mathbf{SL}_n$. Since $\dim(\mathbf{SL}_n \cap \mathbf{D}_n) = n - 1$, it follows rank$(\mathbf{SL}_n) = n - 1$.

5.1.2 Lie Algebra of an Algebraic Group

Given a linear algebraic group $\mathbf{G}$, we denote by Lie$(\mathbf{G})$ the Lie algebra of $\mathbf{G}$. We may think of it as the tangent space $T_e\mathbf{G}$ at $e \in \mathbf{G}$. The definition of the tangent space of an affine variety at a given point can be found in [72, Section 4.1.3]. The vector space isomorphism between Lie$(\mathbf{G})$ and $T_e\mathbf{G}$ is given in [72, Proposition 4.4.5].

Lie Algebras

A **Lie algebra** $\mathfrak{g}$ is a vector space with a bilinear multiplication $[x, y]$ such that $[x, x] = 0$ and such that the **Jacobi identity** holds

$$[[x, y], z] + [[y, z], x] + [[z, x], y] = 0.$$

An important example is the **general linear algebra** $\mathfrak{gl}_n$, which is the vector space $\mathbf{M}_n$ of all $n \times n$ matrices, with the bracket operation $[g, h] = gh - hg$.

Lie Algebra of an Algebraic Group

A **derivation** D of a ring R is a mapping $D : R \to R$ which is linear and satisfies the ordinary rule for derivatives, i.e.,

$$D(x + y) = Dx + Dy \quad \text{and} \quad D(xy) = xDy + yDx.$$

Let $\mathbf{G}$ be an algebraic group and $A = k[\mathbf{G}]$. Let Der A denote the set of all derivations of A. Then Der A is a Lie algebra with respect to the bracket operation $[x, y] = xy - yx$ (Exercises 5.8).

Recall that for $x \in \mathbf{G}$, the left translation L_x and the right translation R_x are given by

$$(L_x f)(y) = f(x^{-1}y) \quad \text{and} \quad (R_x f)(y) = f(yx),$$

where f is a function on G and $y \in G$. Define

$$\mathrm{Lie}(\mathbf{G}) = \{\delta \in \mathrm{Der}\, A \mid \delta L_x = L_x \delta, \forall x \in \mathbf{G}\}.$$

Then $\mathrm{Lie}(\mathbf{G})$ is a Lie algebra (Exercises 5.8). We call $\mathrm{Lie}(\mathbf{G})$ the **Lie algebra of the algebraic group G** (see [38], Section 9.1).

Exercise 5.8 Let **G** be an algebraic group and $A = k[\mathbf{G}]$. Prove that $\mathrm{Der}\, A$ is a Lie algebra, with respect to the bracket operation $[x, y] = xy - yx$. Prove that $\mathrm{Lie}(\mathbf{G})$ is a Lie subalgebra.

Let $g \in \mathbf{G}$. Define the action $\mathrm{Ad}\, g$ on $\mathfrak{g} = \mathrm{Lie}(\mathbf{G})$ by $\mathrm{Ad}\, g(\delta) = R_g \delta R_g^{-1}$. Then $\mathrm{Ad}\, g \in \mathrm{Aut}(\mathfrak{g})$. We define

$$\mathrm{Ad} : G \to \mathrm{Aut}(\mathfrak{g}) \quad \text{by} \quad \mathrm{Ad}(g) = \mathrm{Ad}\, g.$$

Then $\mathrm{Ad} : G \to \mathrm{Aut}(\mathfrak{g})$ is a morphism of algebraic groups, called the **adjoint representation** [72, Proposition 4.4.5].

Example 5.9 Let $\mathbf{G} = \mathbf{GL}_n$, with the Lie algebra $\mathfrak{g} = \mathfrak{gl}_n$. If $g \in \mathbf{G}$, then $\mathrm{Ad}\, g$ acts on $X \in \mathfrak{g}$ by

$$\mathrm{Ad}\, g(X) = gXg^{-1}. \tag{5.1}$$

5.2 Reductive Groups Over Algebraically Closed Fields

Let **G** be a linear algebraic group. The **radical** of **G**, denoted by $R(\mathbf{G})$, is the largest closed connected normal solvable subgroup of **G**. The **semisimple rank** of **G**, denoted by $\mathrm{rank}_{ss}\, \mathbf{G}$, is the rank of $\mathbf{G}/R(\mathbf{G})$.

The subgroup of $R(\mathbf{G})$ consisting of all its unipotent elements is normal in **G**; we call it the **unipotent radical** and denote it by $R_u(\mathbf{G})$. Then $R_u(\mathbf{G})$ is the largest closed connected normal unipotent subgroup of **G**.

A connected algebraic group $\mathbf{G} \neq 1$ is called **semisimple** if $R(\mathbf{G})$ is trivial. A connected algebraic group $\mathbf{G} \neq 1$ is called **reductive** if $R_u(\mathbf{G})$ is trivial.

Example 5.10

(a) Any semisimple group is reductive.
(b) Let $\mathbf{T}$ be a torus. Then $\mathbf{T}$ is solvable, so $R(\mathbf{T}) = \mathbf{T}$. Since all elements of $\mathbf{T}$ are semisimple, the Jordan decomposition implies that the only unipotent element in $\mathbf{T}$ is 1. It follows that $R_u(\mathbf{T}) = 1$, and hence $\mathbf{T}$ is reductive.

Proposition 5.11 *Let* $\mathbf{G}$ *be a reductive group.*

(i) $R(\mathbf{G})$ *is a torus and* $R(\mathbf{G}) = Z(\mathbf{G})^0$, *the identity component of the center of* $\mathbf{G}$.
(ii) The derived subgroup $(\mathbf{G}, \mathbf{G})$ *is semisimple.*
(iii) $\mathbf{G} = R(\mathbf{G})(\mathbf{G}, \mathbf{G})$.
(iv) $\mathrm{rank}_{ss}\, \mathbf{G} = \mathrm{rank}(\mathbf{G}, \mathbf{G})$.
(v) The intersection $R(\mathbf{G}) \cap (\mathbf{G}, \mathbf{G})$ *is finite.*

Proof (i) and (v) follow from [38], page 125. (ii) and (iii) are Corollary 8.1.6 in [72]. (iv) follows from (iii). □

Example 5.12

(a) The groups $\mathbf{SL}_n$, $\mathbf{Sp}_{2n}$, and $\mathbf{SO}_n$ are semisimple, because each of them is equal to its derived subgroup.
(b) The group $\mathbf{G} = \mathbf{GL}_n$ is reductive. Its derived subgroup is $\mathbf{SL}_n$. The radical $R(\mathbf{G})$ consists of all scalar matrices in $\mathbf{G}$.

Theorem 5.13 *Assume that* $\mathbf{G}$ *is connected, semisimple, of rank one. Then* $\mathbf{G}$ *is isomorphic either to* $\mathbf{SL}_2$ *or the projective group* $\mathbf{PSL}_2$.

Proof [72], Theorem 7.2.4. □

5.2.1 Rational Characters

A **rational character** or simply a **character** of a linear algebraic group $\mathbf{G}$ is any morphism of algebraic groups $\chi : \mathbf{G} \to \mathbf{G}_m$. We denote by $X(\mathbf{G})$ the set of all characters of $\mathbf{G}$. It has a natural structure of an abelian group. It $\chi_1, \chi_2 \in X(\mathbf{G})$, we define the product $\chi_1 \chi_2$ by

$$(\chi_1 \chi_2)(g) = \chi_1(g)\chi_2(g).$$

Example 5.14 $\det : \mathbf{GL}_n \to \mathbf{G}_m$ is a character.

Example 5.15 To describe $X(\mathbf{G}_m)$, we first recall that we consider $\mathbf{G}_m$ as a closed subset of $\mathbf{A}^2$ consisting of pairs (x, x^{-1}). Let $\chi : \mathbf{G}_m \to \mathbf{G}_m$ be a character. Then χ is a polynomial in x and x^{-1}, and also a group homomorphism. Hence, it must be of the form

$$\chi(x) = x^n,$$

for some $n \in \mathbb{Z}$. If ψ is another character, $\psi(x) = x^\ell$, then $\chi\psi(x) = x^{n+\ell}$. It follows $X(\mathbf{G}_m) \cong \mathbb{Z}$.

Example 5.16 Since $\mathbf{D}_n \cong \mathbf{G}_m \times \cdots \times \mathbf{G}_m$ (n copies), it follows from the previous example that $X(\mathbf{D}_n) \cong \mathbb{Z}^n$. The character χ of $\mathbf{D}_n$ corresponding to $(m_1, \ldots, m_n) \in \mathbb{Z}^n$ is given by

$$\chi(\mathrm{diag}(a_1, \ldots, a_n)) = a_1^{m_1} a_2^{m_2} \cdots a_n^{m_n}.$$

Let $e_1, \ldots, e_n$ be the standard basis of $\mathbb{Z}^n$. Then, under the above isomorphism $X(\mathbf{D}_n) \cong \mathbb{Z}^n$, we identify e_i with the character of $\mathbf{D}_n$ given by

$$e_i(\mathrm{diag}(a_1, \ldots, a_n)) = a_i.$$

5.2.2   Roots of a Reductive Group

Let $\mathbf{G}$ be a reductive group and $\mathbf{T}$ a maximal torus in $\mathbf{G}$. Let $\mathfrak{g}$ be the Lie algebra of $\mathbf{G}$. We consider the adjoint action of $\mathbf{T}$ on $\mathfrak{g}$. For $\alpha \in X(\mathbf{T})$, define

$$\mathfrak{g}_\alpha = \{X \in \mathfrak{g} \mid \mathrm{Ad}\, t\, X = \alpha(t) X, \forall t \in \mathbf{T}\}.$$

Here, we identify $X(\mathbf{T})$ with $\mathbb{Z}^m$. If $\mathfrak{g}_\alpha \neq 0$ and $\alpha \neq 0$, we call α a **root** of $\mathbf{G}$ relative to $\mathbf{T}$ and $\mathfrak{g}_\alpha$ a **root space**. The set of roots is denoted by $\Phi(\mathbf{G}, \mathbf{T})$ or just Φ and it is called the root system of $\mathbf{G}$ with respect to $\mathbf{T}$. Corresponding to $\alpha = 0$ is the fixed point space

$$\mathfrak{g}^{\mathbf{T}} = \{X \in \mathfrak{g} \mid \mathrm{Ad}\, t\, X = X, \forall t \in \mathbf{T}\}.$$

It is equal to $\mathfrak{t} = \mathrm{Lie}(\mathbf{T})$, the Lie algebra of $\mathbf{T}$. The following is Corollary B(b) in Section 26.2 of [38].

Proposition 5.17 (Root Space Decomposition) *We have*

$$\mathfrak{g} = \mathfrak{t} \oplus \bigoplus_{\alpha \in \Phi} \mathfrak{g}_\alpha,$$

where $\dim \mathfrak{g}_\alpha = 1$ *for all* $\alpha \in \Phi$. *In particular, the set of roots* Φ *is a finite subset of* $X(\mathbf{T})$.

Exercise 5.18 Let $\mathbf{G} = \mathbf{GL}_n$ and $\mathbf{T} = \mathbf{D}_n$. Let $\mathfrak{g} = \mathfrak{gl}_n$, and let $\mathfrak{t}$ be the subalgebra consisting of diagonal elements in $\mathfrak{g}$. The group $\mathbf{T}$ acts on $\mathfrak{g}$ by the adjoint action given by Eq. (5.1). For $i \neq j$, denote by E_{ij} the $n \times n$ matrix having coefficient 1 on the place (i, j), and all other coefficients zero. Let $t = \mathrm{diag}(a_1, \ldots, a_n) \in \mathbf{T}$. Prove that

$$(\mathrm{Ad}\, t)E_{ij} = a_i a_j^{-1} E_{ij}. \tag{5.2}$$

It follows that the character α of $\mathbf{T}$ defined by $\alpha(\mathrm{diag}(a_1, \ldots, a_n)) = a_i a_j^{-1}$ is a root of $\mathbf{G}$. If we denote by $[E_{ij}]$ the linear span of E_{ij}, then (5.2) implies $[E_{ij}] \subseteq \mathfrak{g}_\alpha$. Then $[E_{ij}] = \mathfrak{g}_\alpha$ because $\dim \mathfrak{g}_\alpha = 1$,

From the previous exercise, we can find the roots of $\mathbf{G} = \mathbf{GL}_n$. The group of characters $X(\mathbf{T}) \cong \mathbb{Z}^n$ is generated by $\{e_i \mid i = 1, \ldots, n\}$, where $e_i(a) = a_i$, for $a = \mathrm{diag}(a_1, \ldots, a_n) \in \mathbf{T}$. Define

$$\alpha_{ij} = e_i - e_j.$$

From (5.2), α_{ij} is a root of $\mathbf{G}$. Since $\mathfrak{g} = \mathfrak{t} \oplus \bigoplus_{i \neq j}[E_{ij}]$, it follows that we have found all the roots, so $\Phi(\mathbf{G}, \mathbf{T}) = \{\alpha_{ij} \mid i, j = 1, \ldots n,\ i \neq j\}$.

Proposition 5.19 *Let $\mathbf{G}$ be a reductive group. Let $\mathbf{T}$ be a maximal torus in $\mathbf{G}$, with Lie algebra $\mathfrak{t} = \mathrm{Lie}(\mathbf{T})$. Let $\Phi = \Phi(\mathbf{G}, \mathbf{T})$. Then*

(i) $-\Phi = \Phi$.
(ii) For $\alpha \in \Phi$, let $\mathbf{T}_\alpha$ be the connected component of $\ker \alpha$. Then $\mathbf{T}_\alpha$ is a subtorus of $\mathbf{T}$ of codimension one. We have $Z(\mathbf{G}) = \bigcap_{\alpha \in \Phi} \mathbf{T}_\alpha$.
(iii) The centralizer $\mathbf{Z}_\alpha = \mathbf{Z}_\mathbf{G}(\mathbf{T}_\alpha)$ is a reductive group of semisimple rank one. The groups $\mathbf{Z}_\alpha$, $\alpha \in \Phi$, generate $\mathbf{G}$.
(iv) Let $\mathfrak{z}_\alpha$ be the Lie algebra of $\mathbf{Z}_\alpha$. Then $\mathfrak{z}_\alpha = \mathfrak{t} \oplus \mathfrak{g}_\alpha \oplus \mathfrak{g}_{-\alpha}$

Proof This is Corollary B in Section 26.2 of [38]. □

Example 5.20 Let us write everything explicitly for $\mathbf{G} = \mathbf{GL}_3$ and $\mathbf{T} = \mathbf{D}_3$. Let $\alpha = e_1 - e_2$. Then

$$\ker \alpha = \{\mathrm{diag}(a, a, b) \mid a, b \in k^\times\}.$$

This is a connected group, so $\mathbf{T}_\alpha = \ker \alpha$. Similarly, for $\beta = e_2 - e_3$, we have

$$\ker \beta = \{\mathrm{diag}(a, b, b) \mid a, b \in k^\times\} = \mathbf{T}_\beta.$$

Then

$$\mathbf{T}_\alpha \cap \mathbf{T}_\beta = \ker \alpha \cap ker\beta = \{\mathrm{diag}(a, a, a) \mid a \in k^\times\} = Z(\mathbf{G}),$$

the center of $\mathbf{G}$. Notice that for any $z = \mathrm{diag}(a, a, a) \in Z(\mathbf{G})$, we have $\alpha_{ij}(z) = 1$, for all roots α_{ij}.

To find $\mathbf{Z}_\alpha = Z_{\mathbf{G}}(\mathbf{T}_\alpha)$, we have to do some linear algebra. Let $z \in \mathbf{Z}_\alpha$. We can write it as a block-matrix

$$z = \begin{pmatrix} A & B \\ C & D \end{pmatrix},$$

where A is 2×2, B is 2×1, C is 1×2, and D is 1×1. For any $t = \text{diag}(a, a, b) \in \mathbf{T}_\alpha$, we have $tz = zt$. It is easy to show that we must have $B = 0$ and $C = 0$, while A and D can be arbitrary. It follows

$$\mathbf{Z}_\alpha = \{\begin{pmatrix} g & 0 \\ 0 & a \end{pmatrix} \mid g \in \mathbf{GL}_2, a \in k^\times\}.$$

The Lie algebra of $\mathbf{Z}_\alpha$ is

$$\mathfrak{z}_\alpha = \{\begin{pmatrix} X & 0 \\ 0 & y \end{pmatrix} \mid X \in \mathfrak{gl}_2, y \in k\}.$$

Notice that

$$\mathfrak{g}_\alpha = \{\begin{pmatrix} 0 & x & 0 \\ 0 & 0 & 0 \\ 0 & 0 & 0 \end{pmatrix} \mid x \in k\} \quad \text{and} \quad \mathfrak{g}_{-\alpha} = \{\begin{pmatrix} 0 & 0 & 0 \\ x & 0 & 0 \\ 0 & 0 & 0 \end{pmatrix} \mid x \in k\}.$$

Then $\mathfrak{z}_\alpha = \mathfrak{t} \oplus \mathfrak{g}_\alpha \oplus \mathfrak{g}_{-\alpha}$.

Exercise 5.21 Let $\mathbf{G} = \mathbf{GL}_n$ and $\mathbf{T} = \mathbf{D}_n$. Take a root $\alpha \in \Phi(\mathbf{G}, \mathbf{T})$. It is of the form $\alpha = e_i - e_j, i \neq j$. Compute the groups $\mathbf{T}_\alpha$ and $\mathbf{Z}_\alpha$, and the Lie algebra $\mathfrak{z}_\alpha$.

Weyl Group

Let $\mathbf{G}$ be a connected reductive algebraic group and $\mathbf{T}$ a maximal torus of $\mathbf{G}$. The **Weyl group** of $\mathbf{G}$ with respect to $\mathbf{T}$ is defined as

$$W(\mathbf{G}, \mathbf{T}) = N_{\mathbf{G}}(\mathbf{T})/Z_{\mathbf{G}}(\mathbf{T}),$$

where $N_{\mathbf{G}}(\mathbf{T})$ is the normalizer of $\mathbf{T}$ in $\mathbf{G}$ and $Z_{\mathbf{G}}(\mathbf{T})$ is the centralizer. By Corollary 3.2.9 in [72], $W(\mathbf{G}, \mathbf{T})$ is finite. Since $\mathbf{G}$ is reductive and $\mathbf{T}$ is maximal, we have $Z_{\mathbf{G}}(\mathbf{T}) = \mathbf{T}$. Since all maximal tori in $\mathbf{G}$ are conjugate, their Weyl groups are isomorphic, and we call such a group the **Weyl group** of $\mathbf{G}$.

The roots of $\mathbf{G}$ relative to $\mathbf{T}$ form an abstract root system, as defined below.

Abstract Root Systems

Let E be a Euclidean space: a finite dimensional real vector space with an inner product $(\,,\,)$. Let $\alpha \in E$. Then $H_\alpha = \{x \in E \mid (x, \alpha) = 0\}$ is the hyperplane orthogonal to α. Define

$$s_\alpha(x) = x - \frac{2(x, \alpha)}{(\alpha, \alpha)}\,\alpha.$$

Then $s_\alpha(\alpha) = -\alpha$ and $s_\alpha(x) = x$ for all $x \in H_\alpha$. It follows that s_α is the reflection with respect to H_α. We define a pairing $\langle\,,\,\rangle$ on E by

$$\langle x, y \rangle = \frac{2(x, y)}{(y, y)}.$$

Then

$$s_\alpha(x) = x - \langle x, \alpha \rangle \alpha.$$

Definition 5.22 A (reduced) **abstract root system** is a subset Φ of E such that

(R1) Φ is finite, spans E and does not contain 0.
(R2)   If $\alpha \in \Phi$, the only multiples of α in Φ are $\pm \alpha$.
(R3)   If $\alpha \in \Phi$, then $s_\alpha(\Phi) \subset \Phi$.
(R4)   If $\alpha, \beta \in \Phi$, then $\langle \alpha, \beta \rangle \in \mathbb{Z}$.

The **rank** of Φ is $\dim E$.

   We will give in Theorem 5.27 a classification of root systems. In the following example, we give explicitly root systems of rank 1 and 2.

Example 5.23 There is only one root system of rank 1: A_1.

$$A_1: \quad -\alpha \longleftarrow \quad\longmapsto\quad \longrightarrow \alpha$$

The (reduced) root systems of rank 2 are $A_1 \times A_1$, A_2, B_2, and G_2.

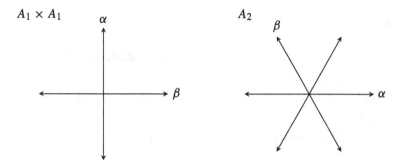

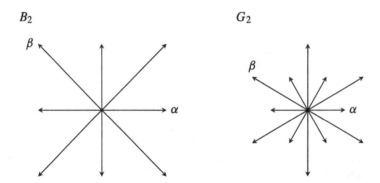

Let Φ be a root system in E. Denote by $W(\Phi)$ the subgroup of $GL(E)$ generated by the reflections s_α, $\alpha \in \Phi$. By (R3), $W(\Phi)$ permutes the set Φ, which is by (R1) finite and spans E. This allows us to identify $W(\Phi)$ with a subgroup of the group of permutations of Φ; in particular, $W(\Phi)$ is finite. We call $W(\Phi)$ the **Weyl group** of Φ.

Theorem 5.24 *Let* **G** *be a reductive group. Let* **T** *be a maximal torus of* **G**, $X = X(\mathbf{T})$ *and* $\Phi = \Phi(\mathbf{G}, \mathbf{T})$. *Denote by Q the subgroup of X generated by Φ. Then,*

(i) Φ *is an abstract root system in* $E = \mathbb{R} \otimes_{\mathbb{Z}} Q$,
(ii) $\operatorname{rank} \Phi = \operatorname{rank}_{ss} \mathbf{G}$,
(iii) $W(\Phi) \cong W(\mathbf{G}, \mathbf{T})$.

Proof If **G** is semisimple, this is Theorem 27.1 in [38]. In general, the theorem follows from Section 7.4 in [72]. □

Example 5.25 Let $\mathbf{G} = \mathbf{GL}_n$ and $\mathbf{T} = \mathbf{D}_n$. Let $\Phi = \Phi(\mathbf{G}, \mathbf{T})$. As explained earlier, $\Phi = \{e_i - e_j \mid i, j = 1, \ldots n, i \neq j\}$. We have $W(\Phi) \cong S_n$, with S_n acting on Φ by permuting $e_1, \ldots, e_n$. The standard inner product on $\mathbb{R} \otimes_{\mathbb{Z}} X(\mathbf{T})$, given by $(e_i, e_j) = \delta_{ij}$, is W-invariant.

The derived subgroup of $\mathbf{G} = \mathbf{GL}_n$ is $\mathbf{SL}_n$. We have $\operatorname{rank} \mathbf{G} = \dim \mathbf{T} = n$ and $\operatorname{rank} \Phi = \operatorname{rank}_{ss} \mathbf{G} = n - 1$. The radical $R(\mathbf{G})$ is equal to the center $Z(\mathbf{G})$ and it consists of all scalar matrices in **G**.

Simple Roots

Let Φ be an abstract root system. A subset Δ of Φ is called a **base** if

(B1) Δ is a basis of E.
(B2)   each root β can be written as $\beta = \sum k_\alpha \alpha$ ($\alpha \in \Delta$) with $k_\alpha \in \mathbb{Z}$ and all $k_\alpha \geq 0$ or all $k_\alpha \leq 0$.

The following is Theorem 10.1 from [39].

Theorem 5.26 *Any abstract root system has a base.*

Fix a base Δ of Φ. From (B1), $|\Delta| = \operatorname{rank} \Phi$. The roots in Δ are called **simple**. If $\alpha \in \Delta$, we call s_α a **simple reflection**.

Let $\beta \in \Phi$ and write $\beta = \sum k_\alpha \alpha$. If all $k_\alpha \geq 0$ (respectively, $k_\alpha \leq 0$), we call β **positive** (respectively, **negative**). We denote by Φ^+ the set of all positive roots and by Φ^- the set of all negative roots.

The Weyl group $W(\Phi)$ is generated by simple reflections and hence any $w \in W(\Phi)$ can be written as a product of simple reflections

$$w = s_1 s_2 \ldots s_k.$$

We call the above decomposition **reduced** if k is as small as possible. If so, k is called the **length of** w relative to Δ and denoted by $\ell(w)$. It turns out that the length of w is equal to the cardinality of the set

$$\{\alpha \in \Phi^+ \mid w(\alpha) \in \Phi^-\}$$

(see Lemma 10.3 A in [39]). In particular, if $w = s_\alpha$ is a simple reflection, then the above set contains just one element, α, and s_α permutes the elements of the set $\Phi^+ - \{\alpha\}$. On the other hand, $W(\Phi)$ contains the unique **longest element**. It is denoted by w_ℓ and it is characterized by the property that $w_\ell(\alpha) < 0$ for every positive root α.

5.2.3 Classification of Irreducible Root Systems

A root system Φ is called **irreducible** if it cannot be partitioned into the union of two proper subsets such that each root in one set is orthogonal to each root in the other. Equivalently, Φ is irreducible if Δ (a base of Φ) cannot be partitioned into the union of two proper subsets such that each root in one set is orthogonal to each root in the other.

Let $\alpha, \beta \in \Phi$. Let θ be the angle between α and β. Then $(\alpha, \beta) = \|\alpha\| \, \|\beta\| \cos \theta$,

$$\langle \beta, \alpha \rangle = \frac{2(\beta, \alpha)}{(\alpha, \alpha)} = 2 \frac{\|\beta\|}{\|\alpha\|} \cos \theta$$

and $\langle \alpha, \beta \rangle \langle \beta, \alpha \rangle = 4 \cos^2 \theta$. This number is an integer, by (R4), and it lays in the interval $[0, 4]$. It follows $\langle \alpha, \beta \rangle \langle \beta, \alpha \rangle = 0, 1, 2$ or 3 (if $\alpha \neq \beta$).

Let $\Delta = \{\alpha_1, \ldots, \alpha_\ell\}$ be a base of Φ. Define the **Coxeter graph** of Φ as the graph with ℓ vertices, with $\langle \alpha_i, \alpha_j \rangle \langle \alpha_j, \alpha_i \rangle$ edges between the ith and the jth vertex ($i \neq j$). Whenever a double or triple edge occurs in the Coxeter graph of Φ, we can add an arrow pointing to the shorter of the two roots. We call the resulting figure the **Dynkin diagram** of Φ. The classification of root systems in terms of

Dynkin diagrams is given by the following two theorems (Theorems 11.4 and 12.1 in Humphreys [39]).

Theorem 5.27 *If Φ is an irreducible system of rank ℓ, its Dynkin diagram is one of the following (ℓ vertices in each case):*

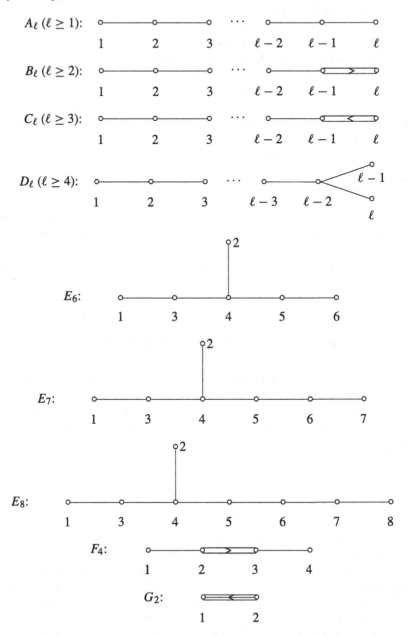

Theorem 5.28 *For each Dynkin diagram of type $A - G$, there exists an irreducible root system having the given diagram.*

5.2.4 Classification of Reductive Groups

A semisimple algebraic group is called **simple** if its root system is irreducible. Such a group has no closed connected normal subgroups other than itself and the trivial group [38, 27.5, 32.1]. The centre $Z(\mathbf{G})$ of a simple group $\mathbf{G}$ is finite, and $\mathbf{G}/Z(\mathbf{G})$ is simple as an abstract group [38, Corollary 29.5].

For each irreducible root system Φ, there exists a simple algebraic group $\mathbf{G}$ having the given root system. The root system Φ, however, does not determine the group $\mathbf{G}$ uniquely—nonisomorphic groups can have the same root system. Already in rank one we have two simple groups, $\mathbf{SL}_2$ and $\mathbf{PSL}_2$, both with root system A_1 (Theorem 5.13). The following is Theorem 32.1 from [38]. It classifies simple algebraic groups in terms of root systems and fundamental groups. (For the definition of the fundamental group of $\mathbf{G}$, see page 127.)

Theorem 5.29 *Suppose $\mathbf{G}$ and $\mathbf{G}'$ are simple algebraic groups over an algebraically closed field k. If they have isomorphic root systems and isomorphic fundamental group, then $\mathbf{G}$ and $\mathbf{G}'$ are isomorphic (as algebraic groups).*

For a classification of reductive groups, we need more data. In addition to characters and roots, we need cocharacters and coroots, as defined below.

Cocharacters

Let $\mathbf{T}$ be a torus, $\mathbf{T} \cong \mathbf{D}_n \cong \mathbf{G}_m \times \cdots \times \mathbf{G}_m$. Any morphism of algebraic groups

$$\lambda : \mathbf{G}_m \to \mathbf{T}$$

is called a **cocharacter** of $\mathbf{T}$. We denote by $X^\vee(\mathbf{T})$ the set of cocharacters of $\mathbf{T}$. It is an abelian group, with the product $(\lambda\mu)(a) = \lambda(a)\mu(a)$. If $\chi \in X(\mathbf{T})$ and $\lambda \in X^\vee(\mathbf{T})$, then

$$\chi \circ \lambda : \mathbf{G}_m \to \mathbf{T} \to \mathbf{G}_m$$

and hence $\chi \circ \lambda \in X(\mathbf{G}_m)$. Using the isomorphism $X(\mathbf{G}_m) \cong \mathbb{Z}$, we can define a natural pairing

$$\begin{aligned} \langle \, , \, \rangle : X(\mathbf{T}) \times X^\vee(\mathbf{T}) &\to \mathbb{Z}, \\ (\chi, \lambda) &\mapsto \langle \chi, \lambda \rangle \in \mathbb{Z} \end{aligned} \tag{5.3}$$

given by $\chi(\lambda(a)) = a^{\langle \chi, \lambda \rangle}$, for all $a \in \mathbf{G}_m$.

Root Datum of a Reductive Group

Let $\mathbf{G}$ be a reductive group and $\mathbf{T}$ a maximal torus of $\mathbf{G}$. We describe here the associated root datum $\Psi(\mathbf{G}, \mathbf{T})$, as outlined in [71]. Details can be found in [72, Chapter 7].

Let $X = X(\mathbf{T})$ and $X^\vee = X^\vee(\mathbf{T})$ (characters and cocharacters). Let $\langle \, , \, \rangle :$ $X(\mathbf{T}) \times X^\vee(\mathbf{T}) \to \mathbb{Z}$ be the natural pairing defined in Eq. (5.3). Take $\Phi = \Phi(\mathbf{G}, \mathbf{T})$, the root system of $\mathbf{G}$ with respect to $\mathbf{T}$. Then Φ is a finite subset of X.

For $\alpha \in \Phi$, we denote by $\mathbf{T}_\alpha$ the identity component of the kernel of α. This is a subtorus of codimension one. Take the centralizer $\mathbf{Z}_\alpha = Z_\mathbf{G}(\mathbf{T}_\alpha)$ and let $\mathbf{G}_\alpha$ be the derived subgroup of $\mathbf{Z}_\alpha$. Then $\mathbf{G}_\alpha$ is a rank one semisimple group, so it is isomorphic to either $\mathbf{SL}_2$ or $\mathbf{PSL}_2$. There is a unique co-character

$$\alpha^\vee : \mathbf{G}_m \to \mathbf{T}$$

such that $\langle \alpha, \alpha^\vee \rangle = 2$, $\operatorname{im} \alpha \subset \mathbf{G}_\alpha$ and $\mathbf{T} = (\operatorname{im} \alpha^\vee) \mathbf{T}_\alpha$. The set $\Phi^\vee = \{\alpha^\vee \mid \alpha \in \Phi\}$ is a finite subset of $X^\vee$. It is also a root system. Its elements are called **coroots**. The quadruple $\Psi(\mathbf{G}, \mathbf{T}) = (X, \Phi, X^\vee, \Phi^\vee)$ is called the root datum of $\mathbf{G}$ with respect to $\mathbf{T}$.

Exercise 5.30 Let $\mathbf{G} = \mathbf{GL}_n$ and $\mathbf{T} = \mathbf{D}_n$. Take a root $\alpha \in \Phi(\mathbf{G}, \mathbf{T})$ of the form $\alpha = e_i - e_{i+1}$, for some $i \in \{1, \dots, n-1\}$. Then $\alpha(\operatorname{diag}(a_1, \dots, a_n)) = a_i a_{i+1}^{-1}$. Compute the groups $\mathbf{T}_\alpha$, $\mathbf{Z}_\alpha$, and $\mathbf{G}_\alpha$ defined above. (The groups $\mathbf{T}_\alpha$ and $\mathbf{Z}_\alpha$ were already computed in Exercise 5.21.) Prove that $\alpha^\vee : \mathbf{G}_m \to \mathbf{T}$ is defined by

$$\alpha^\vee(a) = \operatorname{diag}(1, \dots, 1, a, a^{-1}, 1, \dots, 1),$$

where the entries a and a^{-1} are at the ith and $(i+1)$th position.

Abstract Root Datum

Definition 5.31 A **root datum** is a quadruple $\Psi = (X, \Phi, X^\vee, \Phi^\vee)$ such that

(i) X and $X^\vee$ are free abelian groups of finite type, in duality by a pairing $X \times X^\vee \to \mathbb{Z}$ denoted by $\langle \, , \, \rangle$,
(ii) Φ and $\Phi^\vee$ are finite subsets of X and $X^\vee$ and there is a bijection

$$\Phi \to \Phi^\vee, \qquad \alpha \mapsto \alpha^\vee.$$

(iii) For $\alpha \in \Phi$ we define endomorphisms s_α and $s_\alpha^\vee$ of X and $X^\vee$, respectively, by

$$s_\alpha(x) = x - \langle x, \alpha^\vee \rangle \alpha,$$
$$s_{\alpha^\vee}(u) = u - \langle \alpha, u \rangle \alpha^\vee.$$

Then the following two axioms are imposed:

(RD1)   For all $\alpha \in \Phi$ we have $\langle \alpha, \alpha^\vee \rangle = 2$;
(RD2)   For all $\alpha \in \Phi$ we have $s_\alpha(\Phi) \subset \Phi$, $s_{\alpha^\vee}(\Phi^\vee) \subset \Phi^\vee$.

Theorem 5.32 *Let k be an algebraically closed field. Let $\Psi = (X, \Phi, X^\vee, \Phi^\vee)$ be a root datum. Then there exists a connected reductive linear algebraic group $\mathbf{G}$ over k with a maximal torus $\mathbf{T}$ such that the root datum $\Psi(\mathbf{G}, \mathbf{T})$ is isomorphic to Ψ. The group $\mathbf{G}$ is unique up to an isomorphism of algebraic groups.*

Proof Follows from Springer [72], Theorems 10.1.1 and 9.6.2 (Existence Theorem and Isomorphism Theorem).   □

5.2.5   Structure of Reductive Groups

In this section, $\mathbf{G}$ is a connected reductive algebraic group, $\mathbf{T}$ is a maximal torus of $\mathbf{G}$, and $\Phi = \Phi(\mathbf{G}, \mathbf{T})$ is the root system of $\mathbf{G}$ with respect to $\mathbf{T}$. Let $W = W(\mathbf{G}, \mathbf{T})$.

Root Subgroups

Let $\mathfrak{g} = \mathrm{Lie}(\mathbf{G})$ be the Lie algebra of $\mathbf{G}$ and $\mathfrak{t} = \mathrm{Lie}(\mathbf{T})$. Recall the root space decomposition

$$\mathfrak{g} = \mathfrak{t} \oplus \bigoplus_{\alpha \in \Phi} \mathfrak{g}_\alpha.$$

From Theorem 26.3 in [38] and Section 8.1 in [72] we have the following.

Theorem 5.33 *Let $\alpha \in \Phi$.*

  (i) *There exists a unique connected $\mathbf{T}$-stable subgroup $\mathbf{U}_\alpha$ of $\mathbf{G}$ having Lie algebra $\mathfrak{g}_\alpha$.*
 (ii) *There exists an isomorphism $x_\alpha : \mathbf{G}_a \to \mathbf{U}_\alpha$ such that for any $t \in \mathbf{T}$ and $a \in \mathbf{G}_m$,*

$$t x_\alpha(a) t^{-1} = x_\alpha(\alpha(t)a).$$

(iii) *For any $w \in W$,*

$$w \mathbf{U}_\alpha w^{-1} = \mathbf{U}_{w(\alpha)}.$$

 (iv) *The group $\mathbf{G}$ is generated by the groups $\mathbf{U}_\alpha$ ($\alpha \in \Phi$), along with $\mathbf{T}$.*
  (v) *If $\mathbf{G}$ is semisimple, it is generated by $\mathbf{U}_\alpha$ ($\alpha \in \Phi$).*

The groups $\mathbf{U}_\alpha$ are called **root subgroups**.

Example 5.34 Let $\mathbf{G} = \mathbf{GL}_n$ and $\mathbf{T} = \mathbf{D}_n$. Let $\alpha = e_i - e_j, i \neq j$, be a root of $\mathbf{G}$. We proved in Exercise 5.18 that $\mathfrak{g}_\alpha = [E_{ij}]$, where E_{ij} is the $n \times n$ matrix having coefficient 1 on the place (i, j), and all other coefficients zero. Then

$$\mathbf{U}_\alpha = 1 + \mathfrak{g}_\alpha = \{1 + cE_{ij} \mid c \in k\}.$$

Borel Subgroups and Parabolic Subgroups

A **Borel subgroup** of $\mathbf{G}$ is a maximal closed connected solvable subgroup. Any two Borel subgroups of $\mathbf{G}$ are conjugate ([38], page 134).

Fix a Borel subgroup $\mathbf{P}$ of $\mathbf{G}$ containing $\mathbf{T}$. There is a bijection between the set of bases of Φ and the set of Borel subgroups of $\mathbf{G}$ containing $\mathbf{T}$ ([38], page 166). Therefore, our choice of $\mathbf{P}$ determines the base Δ of Φ and also the set of positive roots Φ^+. We have the decomposition

$$\mathbf{P} = \mathbf{TU}, \quad \mathbf{U} = \prod_{\alpha \in \Phi^+} \mathbf{U}_\alpha.$$

The group $\mathbf{U}$ is the unipotent radical of $\mathbf{P}$. The Borel supgroup opposite $\mathbf{P}$ is

$$\mathbf{P}^- = \mathbf{TU}^-, \quad \mathbf{U}^- = \prod_{\alpha \in \Phi^-} \mathbf{U}_\alpha.$$

Recall that w_ℓ denotes the longest element in W, characterized by the property $w_\ell(\Phi^+) = \Phi^-$. Then $w_\ell \mathbf{U} w_\ell^{-1} = \mathbf{U}^-$ and $w_\ell \mathbf{U}^- w_\ell^{-1} = \mathbf{U}$.

For $w \in W = N_\mathbf{G}(\mathbf{T})/\mathbf{T}$, take a representative $\dot{w}$ in $\mathbf{G}$. Since $\mathbf{T} \subset \mathbf{P}$, the double coset $\mathbf{P}\dot{w}\mathbf{P}$ does not depend on the choice of a representative, and we write simply $\mathbf{P}w\mathbf{P}$. Then we have the following theorem (see Theorem 8.3.8 in [72]):

Theorem 5.35 (Bruhat Decomposition) $\mathbf{G}$ *is the disjoint union*

$$\mathbf{G} = \coprod_{w \in W} \mathbf{P}w\mathbf{P}.$$

Related to the Bruhat decomposition is a partial order on W. For $x, y \in W$, we define

$$x \leq y \quad \text{if} \quad \mathbf{P}x\mathbf{P} \subseteq \overline{\mathbf{P}y\mathbf{P}},$$

where $\overline{\mathbf{P}y\mathbf{P}}$ denotes the closure of $\mathbf{P}y\mathbf{P}$ in $\mathbf{G}$. This partial order can also be defined using the reduced decomposition of the elements of W into product of simple reflections, and it is called **Bruhat order**. More specifically, if $w = s_1 s_2 \ldots s_k$ is a reduced decomposition of $w \in W$, then $x \leq w$ if and only if x can be written as

$x = s_{i_1} s_{i_2} \ldots s_{i_m}$ with $m \geq 0$ and $1 \leq i_1 < \cdots < i_m \leq k$ (see Proposition 8.5.5 in [72]). The longest element w_ℓ is maximal with respect to Bruhat order.

Directly from our definition of the partial order on W, we have

$$\overline{\mathbf{P}w\mathbf{P}} = \coprod_{x \leq w} \mathbf{P}x\mathbf{P}.$$

Since $w \leq w_\ell$ for all $w \in W$, it follows that the closure of $\mathbf{P}w_\ell\mathbf{P}$ is the whole group $\mathbf{G}$. In other words, $\mathbf{P}w_\ell\mathbf{P}$ is dense in $\mathbf{G}$.

Lemma 5.36 *Let $w \in W$ and fix a representative $\dot{w} \in \mathbf{G}$.*

  (i) Every $g \in \mathbf{P}w\mathbf{P}$ can be written in a unique way as

$$g = u\dot{w}p, \quad u \in \mathbf{U} \cap w\mathbf{U}^- w^{-1}, \; p \in \mathbf{P}.$$

  (ii) The map $(u, p) \mapsto u\dot{w}p$ is an isomorphism of varieties $(\mathbf{U} \cap w\mathbf{U}^- w^{-1}) \times \mathbf{P} \cong \mathbf{P}w\mathbf{P}$.

  (iii) $\dim(\mathbf{P}w\mathbf{P}) = \ell(w) + \dim \mathbf{P}$.

Proof Follows from [38, Theorem 28.4] and [72, Lemma 8.3.6]. □

For $w = w_\ell$, we have $w_\ell \mathbf{U}^- w_\ell^{-1} = \mathbf{U}$. Then by Lemma 5.36 (ii) we have an isomorphism $\mathbf{U} \times \mathbf{P} \cong \mathbf{U}w_\ell\mathbf{P} = \mathbf{P}w_\ell\mathbf{P}$. Conjugating by w_ℓ, we see that multiplication induces an isomorphism of varieties

$$\mathbf{U}^- \times \mathbf{P} \xrightarrow{\sim} \mathbf{U}^-\mathbf{P}.$$

The set $\mathbf{U}^-\mathbf{P}$ is open and dense in $\mathbf{G}$ [38, 28.5]. It is called the **big cell**.

A closed subgroup $\mathbf{Q}$ of $\mathbf{G}$ is called **parabolic** if $\mathbf{G}/\mathbf{Q}$ is a projective variety. A closed subgroup of $\mathbf{G}$ is parabolic if and only if it includes a Borel subgroup ([38], page 135). A Borel subgroup is a minimal parabolic subgroup.

5.3   *F*-Reductive Groups

So far in this chapter, we have worked with an algebraically closed field k. Now, we consider an arbitrary subfield F of k.

A closed set V in $\mathbf{A}^n$ is said to be *F*-**closed** if V is the set of zeros of some collection of polynomials having coefficients in F. If $\mathcal{I}(V)$ is generated by F-polynomials, we say that V is **defined over** F. If V is F-closed in $\mathbf{A}^n$, then V is defined over a finite, purely inseparable extension of F (Humphreys [38], Lemma 34.1). When F is perfect (e.g., of characteristic 0), the two notions coincide.

If V is a closed subset of $\mathbf{A}^n$ defined over F, the set

$$V(F) = V \cap F^n$$

is called the set of F-**rational points** of V. In the affine case, when $U \subset \mathbf{A}^n$ and $V \subset \mathbf{A}^m$ are defined over F, we say that a morphism $\varphi : U \to V$ is **defined over** F (or is an F-**morphism**) if the coordinate functions all lie in $F[x_1, \ldots, x_n]$.

Let $\mathbf{G}$ be an algebraic group. If $\mathbf{G}$, along with $\mu : \mathbf{G} \times \mathbf{G} \to \mathbf{G}$ and $\iota : \mathbf{G} \to \mathbf{G}$, are defined over F, we say that $\mathbf{G}$ is **defined over** F or is an F-**group**. Set

$$G = \mathbf{G}(F).$$

A torus $\mathbf{T}$ defined over F is called an F-**torus**. We call $\mathbf{T}$ F-**split** if $\mathbf{T}$ is F-isomorphic to the product of r-copies $\mathbf{G}_m \times \cdots \times \mathbf{G}_m$, where $r = \dim \mathbf{T} = \operatorname{rank} \mathbf{G}$. If so, then

$$\mathbf{T}(F) \cong F^\times \times \cdots \times F^\times \quad (r \text{ copies}).$$

Let $\mathbf{G}$ be a connected reductive F-group. It is said to be **quasi-split** if it contains a Borel subgroup which is defined over F. We say that $\mathbf{G}$ is F-**split** if it has a maximal torus $\mathbf{T}$ which is defined over F and F-split. An example of an F-split group is $\mathbf{GL}_n$ (see page 111).

Let $\mathbf{G}$ be a reductive F-group and $G = \mathbf{G}(F)$. For the structure theory of G, see Borel [8, Capter V] and Borel-Tits [9]. We will work with F-split groups, and for them, the theory is simpler. Here, we just mention some of the objects associated to G in the more general context of reductive F-groups.

The role of a maximal torus is replaced by a maximal F-split torus. Let $\mathbf{T}$ be a maximal F-split torus. Its dimension is called the F-**rank** of $\mathbf{G}$. Define

$$\Phi_F = \Phi(\mathbf{G}, \mathbf{T}) \quad \text{and} \quad W_F = N_{\mathbf{G}}(\mathbf{T})/Z_{\mathbf{G}}(\mathbf{T}).$$

The elements of Φ_F are the roots of $\mathbf{G}$ relative to $\mathbf{T}$ and they are called the F-**roots** of G. The group W_F is called the **Weyl group of G relative to F** (see [72], Section 15.3). The set Φ_F is a root system, but not necessarily reduced. Given $\alpha \in \Phi_F$, it is possible to have $2\alpha \in \Phi_F$ (see [8, Remark 21.7]).

By a parabolic subgroup of $G = \mathbf{G}(F)$, we mean the group of F-rational points $Q = \mathbf{Q}(F)$ of a parabolic subgroup $\mathbf{Q}$ defined over F. The group $\mathbf{G}$ may not posses a Borel subgroup defined over F. In the general theory of F-reductive groups, Borel subgroups are replaced by minimal parabolic subgroups. Let $\mathbf{P}$ be a minimal parabolic subgroup of $\mathbf{G}$ defined over F and $P = \mathbf{P}(F)$. Then, from [8, Theorem 21.15], we have the Bruhat decomposition

$$G = \coprod_{w \in W_F} P w P.$$

Assume $\mathbf{G}$ is F-split. Then $\Phi_F = \Phi$, $W_F = W$, and the structure theory for $\mathbf{G}$ transfers to G. In particular, for $\alpha \in \Phi$, the root group $\mathbf{U}_\alpha$ is defined over F, as well as the isomorphism $x_\alpha : \mathbf{G}_a \to \mathbf{U}_\alpha$. We use the same symbol for the corresponding isomorphism

$$x_\alpha : F \to U_\alpha = \mathbf{U}_a(F).$$

We follow the same convention with roots and coroots of $\mathbf{T}$, and more generally for characters and cocharacters. Corresponding to $\lambda \in X(\mathbf{T})$ is the character of T which we denote by the same letter λ. For a root α, we have

$$\alpha : T \to F^\times \quad \text{and} \quad \alpha^\vee : F^\times \to T.$$

A classification of connected reductive split F-groups by root data can be found in Section 16.3 of [72]. Similarly to Theorem 5.32, there are versions over F of the Isomorphism Theorem and the Existence Theorem [72, Theorems 16.3.2 and 16.3.3].

5.4 Z-Groups

We start with the example of general linear groups. So far, we considered $\mathbf{GL}_n(k)$ and $\mathbf{GL}_n(F)$, but general linear groups are also defined over rings. For a commutative ring with identity R, we define $\mathbf{GL}_n(R)$ as the group of $n \times n$ matrices with coefficients in R and determinant in $R^\times$.

Let L be a finite extension of $\mathbb{Q}_p$, with the ring of integers o_L and the field of fractions $\mathbb{F}_q$. Then we have the following groups

$$\begin{aligned} &\mathbf{GL}_n(\mathbb{Q}), \ \mathbf{GL}_n(\mathbb{Q}_p), \ \mathbf{GL}_n(L), \ \mathbf{GL}_n(\mathbb{Z}), \\ &\quad \mathbf{GL}_n(\mathbb{Z}_p), \ \mathbf{GL}_n(o_L), \ \mathbf{GL}_n(\mathbb{F}_p), \ \mathbf{GL}_n(\mathbb{F}_q). \end{aligned} \tag{5.4}$$

All these groups can be treated simultaneously with the notion of a $\mathbb{Z}$-group.

Let us look back at the description of $\mathbf{GL}_n$ on page 92. All the polynomials defining $\mathbf{GL}_n$ as an algebraic group have coefficients 1 and -1. Hence, they are defined over any subfield F of k. Moreover, they are defined over $\mathbb{Z}$.

Let $\mathbf{T} = \mathbf{D}_n$ be the subgroup of diagonal matrices in $\mathbf{GL}_n$. This is a maximal torus in $\mathbf{GL}_n$. It is defined as the set of zeros of the polynomials x_{ij}, $i \neq j$, and these polynomials are defined over F and also over $\mathbb{Z}$. The torus $\mathbf{T}$ is F-split and we have

$$\mathbf{T}(F) = \mathbf{D}_n(F) \cong F^\times \times \cdots \times F^\times.$$

Having a maximal F-split torus implies that $\mathbf{GL}_n$ is F-split. The $\mathbb{Z}$-points of $\mathbf{T}$ are all diagonal matrices $t = \mathrm{diag}(a_1, \ldots, a_n)$ with coefficients in $\mathbb{Z}$ such that $\det t \in$

$\mathbb{Z}^{\times} = \{\pm 1\}$. It follows

$$\mathbf{T}(\mathbb{Z}) = \{\mathrm{diag}(a_1, \ldots, a_n) \mid a_i \in \{\pm 1\}\} \cong \mathbf{GL}_1(\mathbb{Z}) \times \cdots \times \mathbf{GL}_1(\mathbb{Z}). \qquad (5.5)$$

Before we can talk about $\mathbb{Z}$-split tori, we have to introduce $\mathbb{Z}$-groups. We just mention that $\mathbf{T}$ being $\mathbb{Z}$-split implies not only (5.5), but a more general property (5.6).

5.4.1   Algebraic R-Groups

Let R be a commutative ring with unity. An R-algebra is assumed to be commutative and associative, with unity.

We refer to Jantzen [40], Sections 1 and 2 in Part I, for the definition of an algebraic R-group, which follows the functorial approach of Demazure and Gabriel [25]. It is important to point out that an algebraic R-group $\mathbf{G}_R$ is a functor from the category of R-algebras to the category of groups. Hence,

$$\mathbf{G}_R : A \mapsto \mathbf{G}_R(A),$$

where A is an R-algebra and $\mathbf{G}_R(A)$ is a group. Below, we give basic examples.

Example 5.37 Let $R = \mathbb{Z}$.

1. The **additive group** over $\mathbb{Z}$ is the $\mathbb{Z}$-group functor $\mathbf{G}_{a,\mathbb{Z}}$ with

$$\mathbf{G}_{a,\mathbb{Z}}(A) = (A, +)$$

   for all $\mathbb{Z}$-algebras A. It is an algebraic $\mathbb{Z}$-group. Its polynomial ring $\mathbb{Z}[\mathbf{G}_{a,\mathbb{Z}}]$ is isomorphic to $\mathbb{Z}[x]$.
2. The **multiplicative group** over $\mathbb{Z}$ is the $\mathbb{Z}$-group functor $\mathbf{G}_{m,\mathbb{Z}}$ with

$$\mathbf{G}_{m,\mathbb{Z}}(A) = A^{\times}$$

   for all $\mathbb{Z}$-algebras A. It is an algebraic $\mathbb{Z}$-group with $\mathbb{Z}[\mathbf{G}_{m,\mathbb{Z}}] \cong \mathbb{Z}[x, x^{-1}]$.
3. Let $\mathbf{GL}_n(A)$ be the group of $n \times n$ invertible matrices over A. Then

$$\mathbf{GL}_{n,\mathbb{Z}} : A \mapsto \mathbf{GL}_n(A).$$

   is a $\mathbb{Z}$-group functor from the category of all $\mathbb{Z}$-algebras to the category of groups. It is an algebraic $\mathbb{Z}$-group. Then for

$$A = \mathbb{Q}, \mathbb{Q}_p, L, \mathbb{Z}, \mathbb{Z}_p, o_L, \mathbb{F}_p, \mathbb{F}_q$$

   we obtain the groups listed in Eq. (5.4).

We will work with algebraic $\mathbb{Z}$-groups, and it will be useful to know that such groups can be realized as groups of matrices. Similarly to Theorem 5.2 over algebraically closed fields, we have the following theorem over $\mathbb{Z}$:

Theorem 5.38 *Let* $\mathbf{H}_{\mathbb{Z}}$ *be an algebraic* $\mathbb{Z}$-*group. Then there is a* $\mathbb{Z}$-*isomorphism of* $\mathbf{H}_{\mathbb{Z}}$ *onto a closed subgroup of some* $\mathbf{GL}_{n,\mathbb{Z}}$.

Proof This is Proposition 13.2 (i) in Expose VI$_B$ of [35] in the special case when $S = \mathrm{Spec}(\mathbb{Z})$.   □

We remark that, by the same reference, Theorem 5.38 holds more generally over regular rings of dimension at most 2.

5.4.2 Split Z-Groups

In this section, we will call a split and connected reductive algebraic $\mathbb{Z}$-group simply a **split** $\mathbb{Z}$-**group**. Some authors call such groups *Chevalley groups* [23], but a Chevalley group is more commonly assumed to be semisimple, as in [19].

Let $\mathbf{G}_{\mathbb{Z}}$ be a split $\mathbb{Z}$-group, with a split maximal torus $\mathbf{T}_{\mathbb{Z}}$. Let $r = \dim \mathbf{T}_{\mathbb{Z}} = \mathrm{rank}\,\mathbf{G}_{\mathbb{Z}}$. Then $\mathbf{T}_{\mathbb{Z}} \cong \mathbf{G}_{m,\mathbb{Z}} \times \cdots \times \mathbf{G}_{m,\mathbb{Z}}$, the product of r copies of the multiplicative group over $\mathbb{Z}$ [40, II.1.1]. For any $\mathbb{Z}$-algebra A,

$$\mathbf{T}_{\mathbb{Z}}(A) \cong \mathbf{G}_{m,\mathbb{Z}}(A) \times \cdots \times \mathbf{G}_{m,\mathbb{Z}}(A) = A^{\times} \times \cdots \times A^{\times}. \tag{5.6}$$

Root Subgroups

Let $\Phi = \Phi(\mathbf{G}_{\mathbb{Z}}, \mathbf{T}_{\mathbb{Z}})$ be the set of roots of $\mathbf{G}_{\mathbb{Z}}$ with respect to $\mathbf{T}_{\mathbb{Z}}$. For each $\alpha \in \Phi$ there is a root homomorphism $x_{\alpha} : \mathbf{G}_{a,\mathbb{Z}} \to \mathbf{G}_{\mathbb{Z}}$ with

$$t x_{\alpha}(a) t^{-1} = x_{\alpha}(\alpha(t)a)$$

for any $\mathbb{Z}$-algebra A and all $t \in \mathbf{T}_{\mathbb{Z}}(A)$, $a \in A$, as described in [40, II.1.2]. The functor

$$\mathbf{U}_{\alpha} : A \mapsto x_{\alpha}(\mathbf{G}_{a,\mathbb{Z}}(A)) = x_{\alpha}(A)$$

is a closed subgroup of $\mathbf{G}_{\mathbb{Z}}$ called the root subgroup of $\mathbf{G}_{\mathbb{Z}}$ corresponding to α.

Similarly to reductive groups over an algebraically closed field, split $\mathbb{Z}$-groups are classified by their root data. The proof of Existence and Isomorphism theorems can be found in [26] or [23]. The following is Proposition 6.4.1 from [23].

Theorem 5.39 *A split* $\mathbb{Z}$-*group is determined up to isomorphism by its associated reduced root datum, and every such root datum arises in this way. Two split* $\mathbb{Z}$-*groups are isomorphic over* $\mathbb{Z}$ *if and only if they are isomorphic over* $\mathbb{C}$.

From now on, we will write simply **G** instead of $\mathbf{G}_{\mathbb{Z}}$.

5.5 The Structure of $\mathbf{G}(L)$

In this section, L is a finite extension of $\mathbb{Q}_p$, with ring of integers o_L and unique maximal ideal $\mathfrak{p}_L$, and **G** is a split connected reductive $\mathbb{Z}$-group.

5.5.1 o_L-*Points of Algebraic* $\mathbb{Z}$-*Groups*

We start with some general observations about algebraic $\mathbb{Z}$-groups. Let **H** be an algebraic $\mathbb{Z}$-group. Define

$$H = \mathbf{H}(L) \quad \text{and} \quad H_0 = \mathbf{H}(o_L).$$

We write pr_n for the canonical map $o_L \to o_L/\mathfrak{p}_L^n$ and also for the induced map $H_0 \to \mathbf{H}(o_L/\mathfrak{p}_L^n)$ for any **H**. The kernel of pr_n in H_0 is denoted H_n. Hence,

$$H_n = \{h \in H_0 \mid h \equiv 1 \mod \mathfrak{p}_L^n\}.$$

Finally, $\mathbf{H}(o_L/\mathfrak{p}_L)$ is denoted $\bar{H}$.

When we consider H_0 and H_n as subgroups of H, we do not work with algebraic subgroups (defined using polynomial equations); we are abandoning the territory of algebraic groups and work with topological groups. There is a natural topology on H and we will describe it here.

First, recall that by Theorem 5.38, there is a $\mathbb{Z}$-isomorphism of **H** onto a closed subgroup of some $\mathbf{GL}_m$ (closed in Zariski topology). The group $\mathbf{GL}_m(L)$ is equipped with the standard topology coming from the norm

$$\|g\| = \max_{i,j} |g_{ij}|,$$

for $g = (g_{ij}) \in \mathbf{GL}_m(L)$. We may consider H as a subgroup of $\mathbf{GL}_m(L)$ and equip it with the subspace topology. Notice that $\mathbf{H} \subseteq \mathbf{GL}_m$ is defined as the set of zeros of some set of polynomials. Consequently, H is closed in $\mathbf{GL}_m(L)$ with respect to the standard topology.

Lemma 5.40 *Let* **H** *be an algebraic* $\mathbb{Z}$*-group.*

(i) *The groups* H_n, $n \in \mathbb{N}$, *are normal subgroups of* H_0, *with* H_0/H_n *finite, and*

$$H_0 \cong \varprojlim_{n \in \mathbb{N}} H_0/H_n.$$

(ii) *The projective limit topology on* H_0 *coincides with the standard topology coming from any embedding* **H** $\subseteq$ **GL**$_m$ *of* $\mathbb{Z}$*-groups.*

(iii) *The groups* H_n, $n \in \mathbb{N}$, *are compact and open, and form a neighborhood basis of 1 in* H_0.

Proof Fix an embedding **H** $\subseteq$ **GL**$_m$ of $\mathbb{Z}$-groups and equip H with the subspace topology coming from the standard topology on **GL**$_m(L)$. As the kernel of pr_n : $H_0 \to \mathbf{H}(o_L/\mathfrak{p}_L^n)$, H_n is normal in H_0. For **G** = **GL**$_m$, the quotients G_0/G_n are clearly finite. It follows that H_0/H_n are finite.

The groups G_n, $n \geq 0$ are compact and open in G, and they form a neighborhood basis of 1 in G. Then $H_n = H \cap G_n$ are compact and open in H, and form a neighborhood basis of 1 in H. The canonical projections $H_0 \to H_0/H_n$ are compatible and give in the projective limit the continuous homomorphism

$$\varphi : H_0 \to \varprojlim_{n \in \mathbb{N}} H_0/H_n.$$

By Corollary 2.20, φ is surjective. It is clearly injective. Also, it is easy to see using Proposition 2.32 that φ is open. It follows that φ is an isomorphism of topological groups, thus identifying the standard topology on H_0 with the projective limit topology.   □

5.5.2 o_L-*Points of Split* $\mathbb{Z}$-*Groups*

In the split connected reductive $\mathbb{Z}$-group **G**, we fix a maximal $\mathbb{Z}$-split torus **T**. We denote by W the Weyl group of **G** relative to **T**. For each $w \in W$ we select a representative $\dot{w} \in \mathbf{G}(\mathbb{Z})$.

Fix a Borel subgroup **P** containing **T**. We write Φ for the roots of **T** in **G** and Φ^+ (respectively, Φ^-) for the set of positive (respectively, negative) roots determined by the choice of **P**. For each $\alpha \in \Phi$, we have the root subgroup **U**$_\alpha$ and the morphism x_α from the additive $\mathbb{Z}$-group **G**$_a$ to **U**$_\alpha$.

Let **U** be the unipotent radical of **P**. Then **P** = **TU** and

$$\mathbf{U} = \prod_{\alpha \in \Phi^+} \mathbf{U}_\alpha.$$

The unipotent radical of the opposite parabolic is denoted by $\mathbf{U}^-$. We have $\mathbf{U}^- = \prod_{\alpha \in \Phi^-} \mathbf{U}_\alpha$.

Lemma 5.41 *For $n > 0$, the product map is a homeomorphism*

$$U_n^- \times T_n \times U_n \xrightarrow{\sim} G_n.$$

Proof Follows from [17, Proposition 1.4.4].   □

A nonzero unital ring is called a **local ring** if it has a unique maximal left ideal (equivalently, a unique maximal right ideal) [42, §19].

Proposition 5.42 *For $n \geq 1$,*

(i) *The group G_n is a pro-p group.*
(ii) *The ring $o_K[[G_n]]$ is local, with the unique maximal ideal*

$$\mathfrak{m}(G_n) = \{\mu \in o_K[[G_n]] \mid \mathrm{aug}(\mu) \in \mathfrak{p}_K\}.$$

(iii) *The group of units of $o_K[[G_n]]$ is*

$$o_K[[G_n]]^\times = o_K[[G_n]] \setminus \mathfrak{m}(G_n) = \{\mu \in o_K[[G_n]] \mid \mathrm{aug}(\mu) \in o_K^\times\}.$$

(iv) *The ideals $\mathfrak{m}(G_n)^\ell$, for $\ell \geq 1$, form a fundamental system of neighborhoods of zero.*
(v) *The ring $o_K[[G_n]]$ has no zero divisors.*

Proof (i) First, we prove the assertion for $\mathbf{G} = \mathbf{GL}_m$. Let M be the additive group of $m \times m$ matrices with coefficients in o_L. Then any element $g \in G_n$ can be written in a unique way as $g = 1 + \varpi_L^n X$ with $X \in M$. Define $\varphi : G_n \to \varpi_L^n M / \varpi_L^{n+1} M$ by

$$\varphi(1 + \varpi_L^n X) = \varpi_L^n X \mod \varpi_L^{n+1}.$$

This is a group homomorphism, with kernel G_{n+1}. It follows

$$G_n / G_{n+1} \cong \varpi_L^n M / \varpi_L^{n+1} M \cong M / \varpi_L M.$$

From this, we can show easily that G_n is a pro-p group.

Now, take a general $\mathbf{G}$. By Theorem 5.39, there is a $\mathbb{Z}$-isomorphism of $\mathbf{G}$ onto a closed subgroup $\mathbf{H}$ of $\mathbf{GL}_m$, for some m. Then, we can identify G_n with H_n. As a subgroup of a pro-p group, H_n is also a pro-p group, proving the statement for G_n.

(ii) and (iv) follow from Proposition 19.7 in [63]. For (iii), we use the property that in a local ring, every element is either a unit or it belongs to the maximal ideal [42, Theorem 19.1 and Proposition 19.2]. Assertion (v) follows from Theorem 4.3 of [2], because G_n has no torsion elements.   □

The Bruhat decomposition holds over any field and so we have

$$G = \coprod_{w \in W} P\dot{w}P \quad \text{and} \quad \bar{G} = \coprod_{w \in W} \bar{P}\dot{w}\bar{P}.$$

For $G_0 = \mathbf{G}(o_L)$, however, the Bruhat decomposition is no longer true and the correct one is obtained by pulling back the decomposition for $\bar{G}$. This will require replacing the Borel subgroup with the standard **Iwahori subgroup** $B = \mathrm{pr}_1^{-1}(\bar{P})$.

Proposition 5.43

(i) The standard Iwahori subgroup B factors as $B = U_1^- T_0 U_0 = U_1^- P_0$.
(ii) G_0 is the disjoint union

$$G_0 = \coprod_{w \in W} B\dot{w}B = \coprod_{w \in W} B\dot{w}P_0.$$

Proof Since G_1 is the kernel of the projection map $\mathrm{pr}_1 : G_0 \to \mathbf{G}(o_L/\mathfrak{p}_L)$ and $\mathrm{pr}_1(P_0) = \bar{P}$, we have

$$B = \mathrm{pr}_1^{-1}(\bar{P}) = G_1 P_0 = U_1^- T_1 U_1 P_0 = U_1^- P_0.$$

Here, we use the decomposition $G_1 = U_1^- T_1 U_1$ from Lemma 5.41. Similarly, for any $w \in W$

$$\mathrm{pr}_1^{-1}(\bar{P}\dot{w}\bar{P}) = B\dot{w}B = B\dot{w}U_1^- P_0 = B\dot{w}P_0.$$

Pulling back the Bruhat decomposition $\bar{G} = \coprod_{w \in W} \bar{P}\dot{w}\bar{P}$, we obtain the desired decomposition of G_0. $\square$

Proposition 5.44 *The group G has the Iwasawa decomposition $G = G_0 P$, which refines as the disjoint union*

$$G = \coprod_{w \in W} B\dot{w}P.$$

Proof See [18, Page 392] or [13, Proposition 4.4.3]. $\square$

The decompositions $G = \coprod_{w \in W} P\dot{w}P$ and $G = \coprod_{w \in W} B\dot{w}P$ may look similar, but they are fundamentally different. The pieces $B\dot{w}P$ are approximately of the same size, while the dimension of $P\dot{w}P$ depends on the length of w (see Lemma 5.36). Also, if $w \neq 1$, the double coset $P\dot{w}P$ is not closed in G (see page 109). From [18, Proposition 1.3], we have the following:

(i) $B\dot{w}B \subseteq \coprod_{x \geq w} P\dot{x}P$;
(ii) $B\dot{w}B \cap P\dot{w}P = P_0\dot{w}U_0$.

5.5.3   Coset Representatives for G/P

We will give a set of coset representatives for G/P as in [3, Section 4.1]. We start by defining some unipotent subgroups of G_0.

For $w \in W$, let $V_w^\pm = \dot{w} U^- \dot{w}^{-1}$. Note that $V_w^\pm$ is the product of all the root subgroups U_α attached to roots α such that $w\alpha < 0$. We define

$$U_{w,\frac{1}{2}}^- = \dot{w}^{-1} B \dot{w} \cap U_0^- = (\dot{w}^{-1} U_0 \dot{w} \cap U_0^-)(\dot{w}^{-1} U_1^- \dot{w} \cap U_0^-),$$

$$V_{w,\frac{1}{2}}^\pm = \dot{w} U_{w,\frac{1}{2}}^- \dot{w}^{-1} = (U_0 \cap \dot{w} U^- \dot{w}^{-1})(U_1^- \cap \dot{w} U^- \dot{w}^{-1}).$$

Then

$$V_{w,1}^\pm \subset V_{w,\frac{1}{2}}^\pm \subset V_{w,0}^\pm.$$

The subscript $\frac{1}{2}$ indicates that $V_{w,\frac{1}{2}}^\pm$ is a mixture of $U_{\alpha,1}$'s and $U_{\alpha,0}$'s, while the superscript $\pm$ indicates that some roots α are positive and some are negative. For $U_{w,\frac{1}{2}}^-$, we have

$$U_1^- \subseteq U_{w,\frac{1}{2}}^- \subseteq U_0^-$$

and

$$U_{w,\frac{1}{2}}^- = \prod_{\substack{\alpha < 0 \\ w\alpha > 0}} U_{\alpha,0} \times \prod_{\substack{\alpha < 0 \\ w\alpha < 0}} U_{\alpha,1}.$$

Proposition 5.45 *The set* $\bigsqcup_{w \in W} \dot{w} U_{w,\frac{1}{2}}^-$ *is a set of coset representatives of* G_0/P_0. *In particular,* $B\dot{w}B = \dot{w} U_{w,\frac{1}{2}}^- P_0 = V_{w,\frac{1}{2}}^\pm \dot{w} P_0$ *and we have the disjoint union decomposition*

$$G_0 = \bigsqcup_{w \in W} \dot{w} U_{w,\frac{1}{2}}^- P_0 = \bigsqcup_{w \in W} V_{w,\frac{1}{2}}^\pm \dot{w} P_0. \tag{5.7}$$

Furthermore,

$$G = \bigsqcup_{w \in W} \dot{w} U_{w,\frac{1}{2}}^- P = \bigsqcup_{w \in W} V_{w,\frac{1}{2}}^\pm \dot{w} P.$$

Proof We know that $G_0 P = G$ and $B = G_1 P_0$. By Lemma 5.36, $\mathbf{G} = \coprod_w (\mathbf{U} \cap w\mathbf{U}^- w^{-1}) w\mathbf{P}$. It follows

$$\bar{G} = \coprod_w (\bar{U} \cap w\bar{U}^- w^{-1}) w\bar{P}.$$

Pulling back, we have

$$G_0 = \coprod_w (U_0 \cap \dot{w} U^- \dot{w}^{-1}) \dot{w} G_1 P_0 = \coprod_w (U_0 \cap \dot{w} U^- \dot{w}^{-1}) \dot{w} U_1^- P_0$$

$$= \coprod_w \dot{w} (\dot{w}^{-1} U_0 \dot{w} \cap U_0^-)(\dot{w}^{-1} U_1^- \dot{w} \cap U_0^-) P_0 = \coprod_w \dot{w} U_{w, \frac{1}{2}}^- P_0.$$

$\square$

5.6 General Linear Groups

Let L be a finite extension of $\mathbb{Q}_p$. Chapters 7 and 8 are based on the structure theory of $G(L)$. In this section, we describe explicitly some of the objects associated to general linear groups.

Let $\mathbf{G} = \mathbf{GL}_r$ and $G = \mathbf{GL}_r(L)$. Let $\mathbf{T} = \mathbf{D}_r$ be the maximal torus consisting of diagonal elements. The group of characters $X(\mathbf{T})$ is isomorphic to $\mathbb{Z}^r$, with basis $\{e_1, \dots, e_r\}$ defined by

$$e_i(\mathrm{diag}(a_1, \dots, a_r)) = a_i.$$

The set of roots $\Phi = \Phi(\mathbf{G}, \mathbf{T})$ is

$$\Phi = \{e_i - e_j \mid i, j = 1, \dots r, \ i \neq j\}.$$

The root system is of type A_{r-1}. We select the simple roots

$$\alpha_1 = e_1 - e_2, \quad \alpha_2 = e_2 - e_3, \ \dots, \quad \alpha_{r-1} = e_{r-1} - e_r.$$

Then the corresponding sets of positive and negative roots are

$$\Phi^+ = \{e_i - e_j \mid i < j\} \quad \text{and} \quad \Phi^- = \{e_i - e_j \mid i > j\}.$$

The Weyl group $W = W(\mathbf{G}, \mathbf{T})$ is isomorphic to the symmetric group S_r. If $w = \sigma \in S_r$, then $w(e_i - e_j) = e_{\sigma(i)} - e_{\sigma(j)}$. For every $w \in W$, we take $\dot{w}$ the permutation matrix in G.

Denote by E_{ij} the $r \times r$ matrix having coefficient 1 on the place (i, j), and all other coefficients zero. Then the root subgroup U_α corresponding to the root

$\alpha = e_i - e_j$ is

$$U_\alpha = \mathbf{U}_\alpha(L) = \{1 + cE_{ij} \mid c \in L\}.$$

The root homomorphism $x_\alpha : L \to U_\alpha$ is given by $x_\alpha(c) = 1 + cE_{ij}$.

Let $\mathbf{P} = \mathbf{TU}$ be the Borel subgroup determined by our choice of simple roots. Then $\mathbf{P}$ is the group of upper triangular matrices and $\mathbf{U}$ is the group of unipotent upper triangular matrices. The opposite $\mathbf{U}^-$ is the group of unipotent lower triangular matrices. For instance, for $\mathbf{GL}_3$ we have

$$U = \left\{ \begin{pmatrix} 1 & x & y \\ 0 & 1 & z \\ 0 & 0 & 1 \end{pmatrix} \mid x, y, z \in L \right\} \quad \text{and} \quad U^- = \left\{ \begin{pmatrix} 1 & 0 & 0 \\ x & 1 & 0 \\ y & z & 1 \end{pmatrix} \mid x, y, z \in L \right\}.$$

For every $n \geq 0$, with $\mathfrak{p}_L^0 = o_L$, we have the groups

$$U_n = \left\{ \begin{pmatrix} 1 & x & y \\ 0 & 1 & z \\ 0 & 0 & 1 \end{pmatrix} \mid x, y, z \in \mathfrak{p}_L^n \right\} \quad \text{and} \quad U_n^- = \left\{ \begin{pmatrix} 1 & 0 & 0 \\ x & 1 & 0 \\ y & z & 1 \end{pmatrix} \mid x, y, z \in \mathfrak{p}_L^n \right\}.$$

For short, we write them as

$$U_n = \begin{pmatrix} 1 & \mathfrak{p}_L^n & \mathfrak{p}_L^n \\ 0 & 1 & \mathfrak{p}_L^n \\ 0 & 0 & 1 \end{pmatrix} \quad \text{and} \quad U_n^- = \begin{pmatrix} 1 & 0 & 0 \\ \mathfrak{p}_L^n & 1 & 0 \\ \mathfrak{p}_L^n & \mathfrak{p}_L^n & 1 \end{pmatrix}.$$

The Iwahori subgroup

$$B = U_1^- T_0 U_0$$

can be written as

$$B = \begin{pmatrix} o_L^\times & o_L & o_L \\ \mathfrak{p}_L & o_L^\times & o_L \\ \mathfrak{p}_L & \mathfrak{p}_L & o_L^\times \end{pmatrix}.$$

If we take a matrix $g \in B$ and reduce it modulo $\mathfrak{p}_L$, we obtain an upper triangular matrix.

The group $G_0 = \mathbf{GL}_r(o_L)$ is a maximal compact subgroup of G. The subgroups

$$G_n = \{g \in G_0 \mid g \equiv 1 \mod \mathfrak{p}_L^n\}$$

are compact, open, and normal in G_0. They form a neighborhood basis of the identity. Each G_0/G_n is finite and

$$G_0 \cong \varprojlim_n G_0/G_n.$$

For $n \geq 1$, we have $G_n = U_n^- T_n U_n$. In the case of $\mathbf{GL}_3$, it can be written as

$$G_n = \begin{pmatrix} 1 + \mathfrak{p}_L^n & \mathfrak{p}_L^n & \mathfrak{p}_L^n \\ \mathfrak{p}_L^n & 1 + \mathfrak{p}_L^n & \mathfrak{p}_L^n \\ \mathfrak{p}_L^n & \mathfrak{p}_L^n & 1 + \mathfrak{p}_L^n \end{pmatrix}, \quad n \geq 1.$$

Chapter 6
Algebraic and Smooth Representations

Throughout the book, K and L are finite extensions of $\mathbb{Q}_p$. In this chapter, F is a field of characteristic zero.

In our study of Banach space representations, we will encounter algebraic and smooth representations. Namely, continuous principal series representations may contain finite dimensional algebraic representations or smooth principal series representations. In this chapter, we review some basic properties of these representations.

Section 6.1 is on algebraic representations. We write down some proofs, to give an insight about the structure of algebraic representations, and refer to Jantzen [40] for the complete theory. In Section 6.2, we present some basics on smooth representations. Such representations are usually studied over complex vector spaces, but we want to consider them as subrepresentations of K-Banach space representations. With that in mind, we present smooth representations over an arbitrary field of characteristic zero.

6.1 Algebraic Representations

In this section, we describe all irreducible algebraic representations of a split reductive group. We start with the definition of an algebraic representation of a linear algebraic group and a brief overview of basic properties. For details, we refer to Section I.2 in Jantzen [40].

© The Author(s), under exclusive license to Springer Nature Switzerland AG 2022
D. Ban, *p-adic Banach Space Representations*, Lecture Notes
in Mathematics 2325, https://doi.org/10.1007/978-3-031-22684-7_6

6.1.1 Definition and Basic Properties

Let F be a field of characteristic zero. If V is an F-vector space, we denote by V_a the F-group functor given by

$$V_a(A) = (V \otimes A, +) \tag{6.1}$$

for each F-algebra A.

Let $\mathbf{H}$ be a linear algebraic F-group. An **algebraic representation** of $\mathbf{H}$ on V or a **H-module structure** on V is an action of $\mathbf{H}$ on the F-functor V_a such that each $\mathbf{H}(A)$ acts on $V_a(A) = V \otimes A$ through A-linear maps. Such a representation induces for each A a group homomorphism $\mathbf{H}(A) \to \mathrm{End}_A(V \otimes A)^\times$. This gives us a homomorphism of group functors $\mathbf{H} \to \mathbf{GL}(V)$. Conversely, any such homomorphism defines a representation of $\mathbf{H}$ on V.

If V and V' are $\mathbf{H}$-modules, we denote by $\mathrm{Hom}_{\mathbf{H}}(V, V')$ the space of all $\mathbf{H}$-equivariant maps between V and V'.

A $\mathbf{H}$-module V is called **simple**, and the corresponding representation is called **irreducible**, if $V \neq 0$ and if V has no $\mathbf{H}$-submodules other than 0 and V. It is called **semi-simple** if it is a direct sum of simple $\mathbf{H}$-modules.

For any $\mathbf{H}$-module V, the sum of all its simple submodules is called the **socle** and denoted by $\mathrm{soc}_{\mathbf{H}} V$. If V is semi-simple, then $V = \mathrm{soc}_{\mathbf{H}} V$. Then,

(1) Each simple $\mathbf{H}$-module is finite-dimensional.
(2) If V is a non-zero $\mathbf{H}$-module, then $\mathrm{soc}_{\mathbf{H}} V \neq 0$.
(3) If $\mathbf{H}$ is diagonalisable, then each $\mathbf{H}$-module is semi-simple.

We call $\mathbf{H}$ **unipotent** if it is isomorphic to a closed subgroup of the group of unipotent uppertriangular matrices. Similarly, we say that $\mathbf{H}$ is **trigonalisable** if it is isomorphic to a closed subgroup of the group of upper triangular matrices in some $\mathbf{GL}_n$. Then, we have [40, page 34]:

(4) $\mathbf{H}$ is trigonalisable if and only if each simple $\mathbf{H}$-module has dimension one.
(5) $\mathbf{H}$ is unipotent if and only if, up to isomorphism, F is the only simple $\mathbf{H}$-module.
(6) $\mathbf{H}$ is unipotent if and only if $V^{\mathbf{H}} \neq 0$ for each $\mathbf{H}$-module $V \neq 0$.

Here, $V^{\mathbf{H}}$ denotes $\mathbf{H}$-fixed points in V.

6.1.2 Classification of Simple Modules of Reductive Groups

Let $\mathbf{G}$ be a split reductive F-group. We fix a maximal F-split torus $\mathbf{T}$ and a Borel subgroup $\mathbf{P}$ containing $\mathbf{T}$. We denote by W the Weyl group of $\mathbf{G}$ relative to $\mathbf{T}$. Then

$$W \cong (N_{\mathbf{G}}(\mathbf{T})/\mathbf{T})(A) \cong N_{\mathbf{G}}(\mathbf{T})(A)/\mathbf{T}(A)$$

for any integral A [40, II.1.4, page 157]. Let Φ be the set of roots of $\mathbf{T}$ in $\mathbf{G}$, with Φ^+ (respectively, Φ^-) the set of positive (respectively, negative) roots determined by the choice of $\mathbf{P}$, and with the set of simple roots Δ. The Weyl group $W(\Phi)$ is generated by simple reflections s_α, $\alpha \in \Delta$. We have $W \cong W(\Phi)$.

Let $\mathbf{U}$ be the unipotent radical of $\mathbf{P}$. Then $\mathbf{P} = \mathbf{TU}$. The opposite parabolic is denoted by $\mathbf{P}^-$ and its unipotent radical by $\mathbf{U}^-$.

Abstract Weights

From the definition of an abstract root system, we know that the set of roots Φ spans a Euclidean space E. The space E carries the inner product $(\,,\,)$ and also the pairing $\langle\,,\,\rangle$ defined by $\langle x, y \rangle = 2(x, y)/(y, y)$. Define

$$\Lambda = \{\lambda \in E \mid \langle \lambda, \alpha \rangle \in \mathbb{Z} \text{ for all } \alpha \in \Phi\}.$$

This is a $\mathbb{Z}$-module. The elements of Λ are called the **weights** of Φ. Notice that $\Phi \subset \Lambda$.

We have fixed the set of simple roots Δ. We say that $\lambda \in \Lambda$ is **dominant** if

$$\langle \lambda, \alpha \rangle \geq 0, \quad \text{for all } \alpha \in \Delta.$$

Similarly, we say that $\lambda \in \Lambda$ is **strongly dominant** if $\langle \lambda, \alpha \rangle > 0$ for all $\alpha \in \Delta$. We denote by Λ^+ the set of all dominant weights. If $\Delta = \{\alpha_1, \ldots, \alpha_\ell\}$, define $\lambda_1, \ldots, \lambda_\ell \in E$ by

$$\langle \lambda_i, \alpha_j \rangle = \delta_{ij}.$$

Then $\lambda_1, \ldots, \lambda_\ell \in \Lambda^+$. They are called the **fundamental dominant weights** (relative to Δ).

Recall that a $\mathbb{Z}$-lattice in E is the $\mathbb{Z}$-span of an $\mathbb{R}$-basis of E.

Lemma 6.1 Λ *is a lattice in* E. *It is spanned by the fundamental dominant weights* $\lambda_1, \ldots, \lambda_\ell$.

Proof Notice that $\{\lambda_1, \ldots, \lambda_\ell\}$ is the dual basis to $\{2\alpha_i/(\alpha_i, \alpha_i) \mid i = 1, \ldots, \ell\}$ (relative to the inner product on E). It follows that $\{\lambda_1, \ldots, \lambda_\ell\}$ is a basis of E. Then any $\lambda \in E$ can be written as

$$\lambda = \sum_{i=1}^{\ell} a_i \lambda_i,$$

where $a_1, \ldots, a_\ell \in \mathbb{R}$. Applying $\langle\,,\alpha_j\rangle$ on the equation above gives $a_j = \langle \lambda, \alpha_j \rangle$. In particular, if $\lambda \in \Lambda$, we have $a_i \in \mathbb{Z}$, for all $i = 1, \ldots, \ell$. Hence, Λ is the $\mathbb{Z}$-span of $\{\lambda_1, \ldots, \lambda_\ell\}$. $\qquad\square$

Example 6.2 Suppose Φ is the root system of type A_2. Then $\Delta = \{\alpha_1, \alpha_2\}$,

$$\Phi^+ = \{\alpha_1, \alpha_2, \alpha_1 + \alpha_2\}, \quad \Phi^- = \{-\alpha_1, -\alpha_2, -\alpha_1 - \alpha_2\}.$$

We can compute the fundamental weights λ_1 and λ_2 directly, using $\langle \lambda_i, \alpha_j \rangle = \delta_{ij}$ (see [39, 13.1]). We get $\lambda_1 = \dfrac{1}{3}(2\alpha_1 + \alpha_2)$ and $\lambda_2 = \dfrac{1}{3}(\alpha_1 + 2\alpha_2)$. Note that $\lambda_1 \perp \alpha_2$ and $\lambda_2 \perp \alpha_1$.

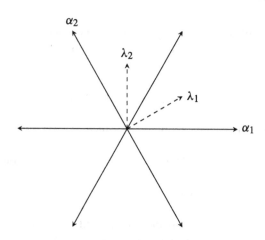

Recall that the Weyl group $W = W(\Phi)$ is generated by reflections s_α, $\alpha \in \Phi$. It preserves the inner product on E and it maps Φ to itself. It follows that W also maps Λ to itself.

We can define a partial order on E as follows:

$$\mu \leq \lambda \iff \lambda - \mu \in \sum_{\alpha \in \Delta} \mathbb{N}\alpha = \sum_{\alpha \in \Phi^+} \mathbb{N}\alpha, \tag{6.2}$$

where $\mathbb{N} = \{0, 1, 2, \dots\}$. This is Lemma A from Humphreys [39, 13.2]:

Lemma 6.3 *Each weight is conjugate under W to one and only one dominant weight. If λ is dominant, then $w\lambda \leq \lambda$ for all $w \in W$, and if λ is strongly dominant, then $w\lambda = \lambda$ only when $w = 1$.*

Denote by Λ_r the $\mathbb{Z}$-span of Φ. Then Λ_r is equal to the $\mathbb{Z}$-span of any base of Φ. It follows that Λ_r is a $\mathbb{Z}$-lattice. It is called the **root lattice**. Since $\Lambda_r \subset \Lambda$, we know from the theory of $\mathbb{Z}$-lattices that Λ/Λ_r is a finite group (called the fundamental group of Φ).

Weights of a Reductive Group

Let $X(\mathbf{T})$ be the group of characters of $\mathbf{T}$. Suppose that $\mathbf{G}$ is semisimple. We know from Theorem 5.24 that Φ is an abstract root system in $E = \mathbb{R} \otimes_{\mathbb{Z}} X(\mathbf{T})$, and rank $\Phi = $ rank $\mathbf{G}$. We have

$$\Lambda_r \subseteq X(\mathbf{T}) \subseteq \Lambda$$

[38, 31.1]. The group $\Lambda / X(\mathbf{T})$ is called the **fundamental group** of $\mathbf{G}$. If $X(\mathbf{T}) = \Lambda$ (so the fundamental group is trivial), we say that $\mathbf{G}$ is **simply connected**. If $X(\mathbf{T}) = \Lambda_r$, we say that $\mathbf{G}$ is **adjoint**.

In general, if we do not assume that $\mathbf{G}$ is semisimple, we know from Proposition 5.11 that the derived subgroup $(\mathbf{G}, \mathbf{G})$ is semisimple and rank $\Phi = $ rank$_{ss}$ $\mathbf{G} = $ rank$(\mathbf{G}, \mathbf{G})$.

We define the set of **dominant weights of** $X(\mathbf{T})$ as

$$X(\mathbf{T})^+ = \{\lambda \in X(\mathbf{T}) \mid \langle \lambda, \alpha^\vee \rangle \geq 0 \text{ for all } \alpha \in \Delta\}$$
$$= \{\lambda \in X(\mathbf{T}) \mid \langle \lambda, \alpha^\vee \rangle \geq 0 \text{ for all } \alpha \in \Phi^+\}.$$

where $\langle \, , \, \rangle : X(\mathbf{T}) \times X^\vee(\mathbf{T}) \to \mathbb{Z}$ is the dual pairing as in (5.3) and $\alpha^\vee$ is the coroot corresponding to α. Similarly as in Eq. (6.2), we define a partial order on $X(\mathbf{T}) \otimes_{\mathbb{Z}} \mathbb{R}$ as follows:

$$\mu \leq \lambda \iff \lambda - \mu \in \sum_{\alpha \in \Delta} \mathbb{N}\alpha = \sum_{\alpha \in \Phi^+} \mathbb{N}\alpha,$$

where $\mathbb{N} = \{0, 1, 2, \dots\}$.

Example 6.4 Let $\mathbf{G} = \mathbf{GL}_n$, with $\mathbf{T}$, Φ, and Δ as in Sect. 5.6. Let $\lambda \in X(\mathbf{T})$. Then $\lambda(\mathrm{diag}(a_1, \dots, a_n)) = a_1^{m_1} \dots a_n^{m_n}$, and $\lambda = m_1 e_1 + \dots + m_n e_n$. For a simple root $\alpha_i = e_i - e_{i+1}$, we have $\langle \lambda, \alpha_i^\vee \rangle = m_i - m_{i+1}$ (see Exercise 5.30). It follows

$$\lambda \in X(\mathbf{T})^+ \iff m_1 \geq m_2 \geq \dots \geq m_n.$$

Dominant Bases of $X(\mathbf{T})$

If $\mathbf{G}$ is simply connected, then the fundamental weights $\lambda_1, \dots, \lambda_r$ belong to $X(\mathbf{T})$ and form a $\mathbb{Z}$-basis of $X(\mathbf{T})$. If $\mathbf{G}$ is not simply connected, we still can find a $\mathbb{Z}$-basis of $X(\mathbf{T})$ which consists of dominant elements. The following is Lemma 2.1 from [3].

Lemma 6.5 *The lattice $X(\mathbf{T})$ has a $\mathbb{Z}$-basis which consists of dominant elements.*

Proof Fix a $\mathbb{Z}$-basis $\underline{\omega} = \{\omega_1, \ldots, \omega_r\}$ of $X(\mathbf{T})$. If $\underline{\lambda} = \{\lambda_1, \ldots, \lambda_r\}$ is a set of linearly independent elements in $X(\mathbf{T})$, let $\mathcal{P}(\underline{\lambda})$ denote the fundamental parallelepiped $\{t_1\lambda_1 + \cdots + t_r\lambda_r : t_1, \ldots, t_r \in [0, 1)\}$. Let $M(\underline{\lambda})$ be the change of basis matrix from $\underline{\omega}$ to $\underline{\lambda}$ and define $d(\underline{\lambda}) = |\det M(\underline{\lambda})|$ (the volume of $\mathcal{P}(\underline{\lambda})$). Then $d(\underline{\lambda}) \in \mathbb{N}$ and $\underline{\lambda}$ is a $\mathbb{Z}$-basis of $X(\mathbf{T})$ if and only if $d(\underline{\lambda}) = 1$.

Write $\mathcal{D}$ for the set of dominant elements in $X(\mathbf{T}) \otimes_{\mathbb{Z}} \mathbb{R}$. Then

1. $\mathcal{D}$ has nonempty interior,
2. $a\mathcal{D} = \mathcal{D}$ for any positive real number a, and
3. $\mathcal{D}$ is convex.

Any subset of $X(\mathbf{T}) \otimes_{\mathbb{Z}} \mathbb{R}$ with the first two properties has the additional property that $\mathcal{D} \cap X(\mathbf{T})$ contains a basis for $X(\mathbf{T}) \otimes_{\mathbb{Z}} \mathbb{R}$. Fix some such basis $\underline{\lambda} = \{\lambda_1, \ldots, \lambda_r\}$. If $\mathcal{P}(\underline{\lambda}) \cap X(\mathbf{T}) = \{0\}$, then $\underline{\lambda}$ is a $\mathbb{Z}$-basis for $X(\mathbf{T})$. Otherwise, take a nontrivial element $\lambda' = t_1\lambda_1 + \cdots + t_r\lambda_r$ of $\mathcal{P}(\underline{\lambda}) \cap X(\mathbf{T})$. Select a nonzero coefficient t_i. Let $\underline{\lambda}'$ be the basis obtained from $\underline{\lambda}$ by replacing λ_i by λ'. As $\mathcal{D}$ is convex, the new basis is still contained in $\mathcal{D} \cap X(\mathbf{T})$. Moreover, $d(\underline{\lambda}') = |t_i|d(\underline{\lambda}) < d(\underline{\lambda})$. After a finite number of steps, we obtain a basis $\underline{\lambda}$ such that $d(\underline{\lambda}) = 1$. $\qquad\square$

Weights of a Module

Suppose that M is a $\mathbf{T}$-module. Since $\mathbf{T}$ is F-split, Proposition 3.2.12 of [72] tells us that M decomposes as a direct sum of weight spaces

$$M = \bigoplus_{\lambda \in X(\mathbf{T})} M_\lambda$$

where $M_\lambda = \{m \in M \mid g(m \otimes 1) = m \otimes \lambda(g) \text{ for all } g \in \mathbf{G}(A) \text{ and all } A\}$ (also, see [40, I.2.10-11]). All λ such that $M_\lambda \neq 0$ are called the **weights** of M.

Notice that the roots of $\mathbf{G}$ with respect to $\mathbf{T}$ are the weights of $\mathfrak{g} = \mathrm{Lie}(\mathbf{G})$ with respect to the adjoint action of $\mathbf{T}$ on $\mathfrak{g}$.

If M is a $\mathbf{G}$-module, then it is of course a $\mathbf{T}$-module, and it decomposes as above into a direct sum of weight spaces. We can look at the action of the Weyl group. Let $\dot{w} \in N_{\mathbf{G}}(\mathbf{T})(F)$ be a representative of $w \in W$. It is easy to show that

$$\dot{w} M_\lambda = M_{w(\lambda)}, \quad \text{for all } \lambda \in X(\mathbf{T}).$$

Lemma 6.6 *If λ is maximal among the weights of M, then $M_\lambda \subseteq M^{\mathbf{U}}$.*

Proof We discuss the proof briefly and refer to [40, II.1.19] for details.

Recall that for each root $\alpha \in \Phi$ we have an isomorphism $x_\alpha : \mathbf{G}_a \to \mathbf{U}_\alpha$. The tangent map dx_α induces an isomorphism $dx_\alpha : \mathrm{Lie}(\mathbf{G}_a) \to \mathfrak{g}_\alpha$. Let $X_\alpha = dx_\alpha(1)$ and $X_{\alpha,n} = X_\alpha^n/(n!) \otimes 1$. Then, it can be shown that

$$X_{\alpha,n} M_\lambda \subset M_{\lambda+n\alpha}.$$

On the other hand, X_α determines the action of $\mathbf{U}_\alpha$ on M: for $m \in M$ and $a \in A$, we have

$$x_\alpha(a)(m \otimes 1) = \sum_{n \geq 0} (X_{\alpha,n} m) \otimes a^n.$$

If λ is maximal among the weights of M, then for any $\alpha \in \Phi^+$ we have $M_{\lambda+n\alpha} = 0$, for any $n > 0$. It follows $X_{\alpha,n} m = 0$ for any $m \in M_\lambda$ and $n > 0$. Hence,

$$x_\alpha(a)(m \otimes 1) = m \otimes 1$$

for any $m \in M$ and $a \in A$, so M_λ is fixed by $\mathbf{U}_\alpha$. This holds for all $\alpha \in \Phi^+$. Consequently, M_λ is fixed by $\prod_{\alpha \in \Phi^+} \mathbf{U}_\alpha = \mathbf{U}$. $\qquad \square$

Algebraic Induction

Let $\mathbf{H}$ be a subgroup scheme of $\mathbf{G}$. Each $\mathbf{G}$-module has a natural structure of an $\mathbf{H}$-module given by the restriction of the action of $\mathbf{G}(A)$ to $\mathbf{H}(A)$ for each F-algebra A. We obtain a functor

$$\mathrm{res}_{\mathbf{H}}^{\mathbf{G}} : \{\mathbf{G}\text{-modules}\} \to \{\mathbf{H}\text{-modules}\}.$$

On the other hand, let M be an $\mathbf{H}$-module. Define

$$\mathrm{ind}_{\mathbf{H}}^{\mathbf{G}}(M) = \{f \in \mathrm{Mor}(\mathbf{G}, M_a) \mid f(gh) = h^{-1} f(g)$$

$$\text{for all } g \in \mathbf{G}(A), h \in \mathbf{H}(A) \text{ and all } A\},$$

where M_a is as in Eq. (6.1). Then $\mathrm{ind}_{\mathbf{H}}^{\mathbf{G}}(M)$ is a $\mathbf{G}$-module, with $\mathbf{G}$ action by left translations. It is called the **induced module** of M from $\mathbf{H}$ to $\mathbf{G}$. Then $M \mapsto \mathrm{ind}_{\mathbf{H}}^{\mathbf{G}}(M)$ defines a functor

$$\mathrm{ind}_{\mathbf{H}}^{\mathbf{G}} : \{\mathbf{H}\text{-modules}\} \to \{\mathbf{G}\text{-modules}\}.$$

For the properties of the functors $\mathrm{res}_{\mathbf{H}}^{\mathbf{G}}$ and $\mathrm{ind}_{\mathbf{H}}^{\mathbf{G}}$, see section I.3 in [40].

We are interested in the parabolic induction. Let $\lambda \in X(\mathbf{T})$. We extend λ trivially to $\mathbf{U}^-$ and obtain a character of $\mathbf{P}^- = \mathbf{T}\mathbf{U}^-$, which we denote again by λ. Then λ defines a one-dimensional $\mathbf{P}^-$-module denoted by $F^{(\lambda)}$ or simply λ. The induced module is

$$\mathrm{ind}_{\mathbf{P}^-}^{\mathbf{G}}(\lambda) = \{f \in F[\mathbf{G}] \mid f(gp) = \lambda(p)^{-1} f(g)$$

$$\text{for all } g \in \mathbf{G}(A), p \in \mathbf{P}^-(A) \text{ and all } A\}.$$

The Frobenius reciprocity [40, Section I.3.4] for $\mathrm{ind}_{\mathbf{P}_-}^{\mathbf{G}}$ gives us

$$\mathrm{Hom}_{\mathbf{G}}(V, \mathrm{ind}_{\mathbf{P}_-}^{\mathbf{G}}(\lambda)) \cong \mathrm{Hom}_{\mathbf{P}_-}(\mathrm{res}_{\mathbf{P}_-}^{\mathbf{G}} V, \lambda) \qquad (6.3)$$

for each $\mathbf{G}$-module V.

Lemma 6.7 *Let $\lambda \in X(\mathbf{T})$. Then $\mathrm{ind}_{\mathbf{P}_-}^{\mathbf{G}}(\lambda) \neq 0$ if and only if λ is dominant.*

Proof Jantzen [40], Proposition II 2.6.   □

Remark 6.8 We may switch the roles of $\mathbf{P}$ and $\mathbf{P}^-$, but we have to be careful about the dominance condition. With the same definition of a dominant weight as before, we proceed as follows. Let $\mu \in X(\mathbf{T})$ and

$$\mathrm{ind}_{\mathbf{P}}^{\mathbf{G}}(\mu) = \{f \in F[\mathbf{G}] \mid f(gp) = \mu(p)^{-1} f(g) \text{ for all } g \in \mathbf{G}(A),\, p \in \mathbf{P}(A) \text{ and all } A\}.$$

Then $\mathrm{ind}_{\mathbf{P}}^{\mathbf{G}}(\mu) \neq 0$ if and only if $\lambda = -\mu$ is dominant.

Suppose $\lambda \in X(\mathbf{T})$ is dominant. Let $V = \mathrm{ind}_{\mathbf{P}_-}^{\mathbf{G}}(\lambda)$. We would like to say more about the structure of V. As a $\mathbf{T}$-module, it decomposes as

$$V = \bigoplus_{\mu \in X(\mathbf{T})} V_\mu.$$

From the Frobenius reciprocity (6.3), we have

$$\mathrm{Hom}_{\mathbf{G}}(V, V) = \mathrm{Hom}_{\mathbf{G}}(V, \mathrm{ind}_{\mathbf{P}_-}^{\mathbf{G}}(\lambda)) \cong \mathrm{Hom}_{\mathbf{P}_-}(\mathrm{res}_{\mathbf{P}_-}^{\mathbf{G}} V, \lambda) \subset \mathrm{Hom}_{\mathbf{T}}(\mathrm{res}_{\mathbf{T}}^{\mathbf{G}} V, \lambda).$$

Then $\mathrm{Hom}_{\mathbf{G}}(V, V) \neq 0$ implies that $\mathrm{Hom}_{\mathbf{T}}(\mathrm{res}_{\mathbf{T}}^{\mathbf{G}} V, \lambda) \neq 0$ as well. It follows that λ is a weight of V.

Lemma 6.9 *Suppose $\lambda \in X(\mathbf{T})$ is dominant, and $V = \mathrm{ind}_{\mathbf{P}_-}^{\mathbf{G}}(\lambda)$. Then*

(i) $\dim V^{\mathbf{U}} = 1$ *and* $V^{\mathbf{U}} = V_\lambda$.
(ii) λ *is the unique maximal weight of V.*
(iii) *Each weight μ of V satisfies $w_\ell \lambda \leq \mu \leq \lambda$, where w_ℓ is the longest element in W.*

Recall that the longest element w_ℓ in W may be characterized by the property that $w_\ell(\alpha) < 0$ for every positive root α, or equivalently $w_\ell(\alpha) < 0$ for every $\alpha \in \Delta$.

Proof By property (6) on page 124, $V^{\mathbf{U}} \neq 0$. If $f \in V^{\mathbf{U}}$, then $u.f = f$ for all $u \in \mathbf{U}$, where the action is by left translation. Then

$$f(utu^-) = \lambda(t)^{-1} f(1),$$

for all $u \in \mathbf{U}(A)$, $t \in \mathbf{T}(A)$, $u^- \in \mathbf{U}^-(A)$, and all A. Hence, $f(x)$ is determined by $f(1)$, for all $x \in \mathbf{UP}^-$. Since $\mathbf{UP}^-$ is dense in $\mathbf{G}$, it follows that $f \in V^{\mathbf{U}}$ is completely determined by $f(1)$. Hence,

$$\dim V^{\mathbf{U}} = 1.$$

The evaluation map $\varepsilon : V \to \lambda$, $f \mapsto f(1)$ is a homomorphism of $\mathbf{P}^-$-modules and is injective on $V^{\mathbf{U}}$. This implies $V^{\mathbf{U}} \subseteq V_\lambda$. Suppose μ is a maximal weight of V. By Lemma 6.6 and the above equation, we have

$$V_\mu \subset V^{\mathbf{U}} \subseteq V_\lambda.$$

It follows $\mu = \lambda$ and $V^{\mathbf{U}} = V_\lambda$.

Now, suppose μ is an arbitrary weight of V. Then μ is less than or equal to a maximal weight. Since V has the unique maximal weight λ, it follows $\mu \leq \lambda$. As observed earlier, $w\mu$ is a weight of V, for any $w \in W$. In particular, $w_\ell\mu$ is a weight of V, and hence $w_\ell\mu \leq \lambda$. By the definition of the partial order on $X(\mathbf{T})$, $\lambda - w_\ell\mu \in \sum_{\alpha\in\Phi^+} \mathbb{N}\alpha$. Then

$$w_\ell\lambda - \mu \in \sum_{\alpha\in\Phi^+} \mathbb{N}(w_\ell\alpha) = \sum_{\beta\in\Phi^-} \mathbb{N}\beta.$$

Since $\Phi^- = -\Phi^+$, it follows $w_\ell\lambda \leq \mu$.   □

Simple Modules

Let V be a simple $\mathbf{G}$-module. By property (1), V is finite-dimensional. Since $\mathbf{U}$ and $\mathbf{U}^-$ are unipotent, from property (6) we have

$$V^{\mathbf{U}} \neq 0 \quad \text{and} \quad V^{\mathbf{U}^-} \neq 0.$$

Since $\mathbf{T}$ normalizes $\mathbf{U}$ and $\mathbf{U}^-$, it follows that $V^{\mathbf{U}}$ and $V^{\mathbf{U}^-}$ are $\mathbf{T}$-submodules of V and so they decompose as direct sums of their weight spaces.

Let λ be a weight of $V^{\mathbf{U}}$. Then $V^{\mathbf{U}}_\lambda$ is also a $\mathbf{P}$-module and by property (4) it decomposes as a direct sum of one-dimensional subspaces. On each of these subspaces, $\mathbf{T}$ acts as λ and $\mathbf{U}$ acts trivially, so each is isomorphic to λ.

The same reasoning applies to $\mathbf{P}^-$. It follows that there exist $\lambda, \mu \in X(\mathbf{T})$ such that

$$\mathrm{Hom}_{\mathbf{P}^-}(V, \lambda) \neq 0 \quad \text{and} \quad \mathrm{Hom}_{\mathbf{P}}(V, \mu) \neq 0.$$

Here, V is considered as a $\mathbf{P}^-$-module (respectively, $\mathbf{P}$-module), and we could also write $\mathrm{res}^{\mathbf{G}}_{\mathbf{P}_-} V$ (respectively, $\mathrm{res}^{\mathbf{G}}_{\mathbf{P}} V$). Then the Frobenius reciprocity (6.3) implies

$$\mathrm{Hom}_{\mathbf{G}}(V, \mathrm{ind}^{\mathbf{G}}_{\mathbf{P}_-}(\lambda)) \neq 0 \quad \text{and} \quad \mathrm{Hom}_{\mathbf{G}}(V, \mathrm{ind}^{\mathbf{G}}_{\mathbf{P}}(\mu)) \neq 0. \tag{6.4}$$

Lemma 6.7 implies that such λ is dominant. If $\lambda \in X(\mathbf{T})$ is dominant, we define

$$M(\lambda) = \mathrm{ind}^{\mathbf{G}}_{\mathbf{P}_-}(\lambda).$$

Proposition 6.10

(i) $M(\lambda)$ is a simple $\mathbf{G}$-module.
(ii) Any simple $\mathbf{G}$-module is isomorphic to exactly one $M(\lambda)$ with λ dominant.

Proof

(i) This follows from [40], II.2.4 and II.5.6. We remark that we assume char $F = 0$. In general, the simple $\mathbf{G}$-module associated to the dominant weight λ is $M(\lambda) = \mathrm{soc}_{\mathbf{G}}\, \mathrm{ind}^{\mathbf{G}}_{\mathbf{P}_-}(\lambda)$.
(ii) Suppose V is a simple $\mathbf{G}$-module. From (6.4), we have $\mathrm{Hom}_{\mathbf{G}}(V, M(\lambda)) \neq 0$, for some $\lambda \in X(\mathbf{T})$. Since V and $M(\lambda)$ are both simple, they must be isomorphic.

$\square$

6.2   Smooth Representations

In this section, G is a locally profinite group. We study smooth (that is, locally constant) representations of G. There is a well-established theory of smooth representations of G on complex vector spaces. We would like to know what parts of the complex theory carry over if we replace $\mathbb{C}$ with a finite extension K of $\mathbb{Q}_p$. There are two obstacles. First, K is not algebraically closed. Second, there is no K-valued Haar measure on G (see Sect. 3.4.2).

We work over an arbitrary field F of characteristic zero. Vignéras in Chapter I of [78] presents a basic theory of representations of a locally profinite group over a commutative ring.

We say that a function $f : G \to F$ is **smooth** if it is locally constant. Denote by $C^\infty_c(G, F)$ the space of smooth compactly supported functions $f : G \to F$. Combining [78, Ch.I, 2.3] and [14, 3.1], we define a left **Haar integral** on G with values in F as a nonzero linear functional

$$I : C^\infty_c(G, F) \to F$$

which is invariant under left translations by G. If S is a compact open subset of G, we define $\mu(S) = I(1_S)$, where 1_S is the characteristic function of S. We also use the integral notation and write

$$\int_G f(x)d\mu(x)$$

for $I(f)$. Notice that μ satisfies properties (i) and (ii) of Definition 3.45, but not necessarily the boundedness condition (iii). This causes no troubles for compactly supported smooth functions. Namely, for $f \in C_c^\infty(G, F)$, the integral $\int_G f(x)d\mu(x)$ is in fact a finite sum and it is well-defined over F, as we can see from the proof of the proposition below.

Proposition 6.11 *There exists a nontrivial left Haar integral on G with values in F. It is unique up to a multiplicative constant.*

Proof Take a nonzero scalar $c \in F$ and define $\mu(G_0) = c$. If $f \in C_c^\infty(G, F)$, then f is compactly supported and locally constant. Hence, there exist a compact open subgroup H of G_0 and a finite number of elements $g_1, \ldots, g_n \in G$ such that $\mathrm{supp}(f) \subseteq \bigcup_{i=1}^n g_i H$ and $f(g_i h) = f(g_i)$ for all $h \in H$ and all i. Define

$$\int_G f(x)d\mu(x) = \sum_{i=1}^n \frac{c}{|G_0 : H|} f(g_i).$$

It is easy to show that this defines a nontrivial left Haar integral on G with values in F. Also, it is easy to show that any Haar integral must be of the same form, so it is unique up to a choice of $c \in F$. $\square$

6.2.1 Absolute Value

The definition of an absolute value on a field is given in Definition A.6 in Appendix A.2.1.

The absolute value on L can be computed as follows (see Appendix A.2.2). Let o_L be the ring of integers of L and $\mathfrak{p}_L$ the unique maximal ideal of o_L. The ideal $\mathfrak{p}_L$ is principal. Let ϖ_L be a generator of $\mathfrak{p}_L$. Denote by q_L order of the residue field $o_L/\mathfrak{p}_L$. Then $|\varpi_L|_L = q_L^{-1}$. Any nonzero $x \in L$ can be written as $x = \varpi_L^m u$, where $m = v_L(x) \in \mathbb{Z}$ and $u \in o_L^\times$. Then

$$|x|_L = |\varpi_L^m u|_L = q_L^{-m}.$$

Notice that for any $x \in L$, $|x|_L$ is a rational number. Since char $F = 0$, we have $\mathbb{Q} \subseteq F$. Then $|x|_L$, which by Definition A.6 is an element of $\mathbb{R}_{\geq 0}$, can also be considered as an element of F. Still, when we write $o_L = \{x \in L \mid |x|_L \leq 1\}$, i.e., if we consider o_L to be the unit ball in L, we take $|x|_L$ to be a real number. On the other hand, if we talk about representations built using $|\ |_L$, then $|x|_L \in F$ (see Example 6.13).

6.2.2 Smooth Representations and Characters

Let V be an F-vector space. Recall that a representation (π, V) of G is a group homomorphism

$$\pi : G \to \mathrm{Aut}(V).$$

A **character** of G is a one-dimensional representation of G. Equivalently, a character χ of G is a group homomorphism

$$\chi : G \to F^{\times}.$$

So far, we talked about Banach space representations (Chap. 4) and algebraic representations (Sect. 6.1). Now, we define smooth representations.

Definition 6.12

(i) A representation (π, V) is called **smooth** if for every $v \in V$, there is a compact open subgroup H of G such that $\pi(h)v = v$ for all $h \in H$.

(ii) A smooth representation (π, V) of G is called **admissible-smooth** if the space of H-fixed vectors V^H is finite-dimensional for every compact open subgroup H of G. Here, $V^H = \{v \in V \mid \pi(h)v = v \text{ for all } h \in H\}$.

Notice that (π, V) is smooth if and only it is continuous with respect to the discrete topology on V. The definition of a smooth representation does not require a topology on V (and does not see it, if such topology exists). Similarly, there are no continuity requirements on intertwining operators: if (π, V) and (τ, W) are smooth representations of G, we denote by $\mathrm{Hom}_G(\pi, \tau)$ or $\mathrm{Hom}_G(V, W)$ the space of G-equivariant linear maps $f : V \to W$. We denote by $\mathrm{Rep}_F^{\mathrm{sm}}(G)$ the category of smooth representations of G on F-vector spaces.

Here are first examples of smooth representations: some smooth characters.

Example 6.13

(a) Let $\mathbf{G} = \mathbf{GL}_1$, so $G = \mathbf{G}(L) = L^{\times}$. Then $|\ |_L : L^{\times} \to F^{\times}$ is a smooth character of $L^{\times}$.

(b) Let $G = \mathbf{GL}_n(L)$. Define $\nu : G \to F^{\times}$ by $\nu(g) = |\det(g)|_L$. Then ν is a smooth character of G.

(c) Let $G = \mathbf{GL}_n(L)$ and let $\chi : G \to F^{\times}$ be an arbitrary character. Let $H = \mathbf{SL}_n(L)$. Then H is the derived subgroup of G, that is, the group generated by all commutators $[g, h] = ghg^{-1}h^{-1}$, $g, h \in G$. Notice that $\chi([g, h]) = \chi(g)\chi(h)\chi(g^{-1})\chi(h^{-1}) = 1$. It follows that χ is trivial on H and hence

$$\chi = \eta \circ \det$$

where $\eta : L^{\times} \to F^{\times}$ is a character of $L^{\times}$. In particular, if χ is smooth, then $\chi = \eta \circ \det$, for a smooth character η.

6.2.3 Basic Properties

Some basic results about smooth representations of G on F-vector spaces, where F is an arbitrary field of characteristic zero, can be found in Section 2 of Casselman [17]. In this section, we present some of these results. Also, cf. Vignéras [78, Ch.I].

We start with a comment on isomorphic fields.

Isomorphic Fields

Let K be a finite extension of $\mathbb{Q}_p$ and let $\overline{\mathbb{Q}}_p$ be an algebraic closure of $\mathbb{Q}_p$ containing K. As we will see below, some results on smooth representations hold only if the coefficient field is algebraically closed. Because of that, it is sometimes useful to work first over $\overline{\mathbb{Q}}_p$, and then deduce results for K.

Hence, we would like to understand the category $\mathrm{Rep}^{\mathrm{sm}}_{\overline{\mathbb{Q}}_p}(G)$. Interestingly, it is equivalent to $\mathrm{Rep}^{\mathrm{sm}}_{\mathbb{C}}(G)$. The reason is because the fields $\overline{\mathbb{Q}}_p$ and $\mathbb{C}$ are isomorphic as abstract fields (see Appendix A.2.3). This may sound contra-intuitive because we usually consider them equipped with standard metrics. However, for smooth representations, the coefficient field is considered as an abstract field.

Proposition 6.14 *If the fields E and F are isomorphic as abstract fields, then the categories $\mathrm{Rep}^{\mathrm{sm}}_E(G)$ and $\mathrm{Rep}^{\mathrm{sm}}_F(G)$ are equivalent.*

Proof We apply restriction of scalars. Fix an isomorphism $\iota : F \to E$. Any E-vector space V becomes an F-vector space via

$$(a, v) \mapsto \iota(a)v, \quad a \in F, v \in V.$$

Also, if $f \in \mathrm{Hom}_E(V, W)$, then clearly $f \in \mathrm{Hom}_F(V, W)$.

Now, take a smooth representation $(\pi, V) \in \mathrm{Rep}^{\mathrm{sm}}_E(G)$. The homomorphism $\pi : G \to \mathrm{Aut}_E(V)$ can be seen as a homomorphism $\pi : G \to \mathrm{Aut}_F(V)$ and therefore $(\pi, V) \in \mathrm{Rep}^{\mathrm{sm}}_F(G)$. It is now easy to show that the resulting map $\mathrm{Rep}^{\mathrm{sm}}_E(G) \to \mathrm{Rep}^{\mathrm{sm}}_F(G)$ is an equivalence of categories. $\square$

Absolutely Irreducible Representations

A representation V is irreducible if V does not contain G-invariant subspaces except zero and itself. Since we are not assuming that F is algebraically closed, we have to distinguish irreducibility from absolute irreducibility.

Let (π, V) be a representation of G on the F-vector space V and let E/F be a field extension. We denote by $\pi \otimes E$ or by π_E the representation of G on the E-vector space $V_E = V \otimes_F E$ given by

$$(\pi \otimes E)(g)(v \otimes c) = \pi(g)v \otimes c$$

and call it the scalar extension of π to E. Some results about the extension of smooth representations from F to E and the descent from E to F can be found in [37].

We say that π is **absolutely irreducible** if $\pi \otimes E$ is irreducible for any field extension E/F.

Lemma 6.15 *Let (π, V) be a representation of G over F and let E/F be a field extension. Then,*

(i) (π, V) is smooth if and only if $(\pi \otimes E, V \otimes_F E)$ is smooth.
(ii) (π, V) is smooth-admissible if and only if $(\pi \otimes E, V \otimes_F E)$ is smooth-admissible.

Proof If H is a subgroup of G, we claim that $(V \otimes E)^H = V^H \otimes E$. Clearly, $V^H \otimes E \subseteq (V \otimes E)^H$. For the converse inclusion, take

$$\mu = \sum_{finite} v_i \otimes c_i \in (V \otimes E)^H.$$

We may assume that c_i are linearly independent over F. By the assumption on μ, for any $h \in H$, we have $(\pi \otimes E)(h)\mu = \mu$, that is,

$$(\pi \otimes E)(h)(\sum v_i \otimes c_i) = \sum \pi(h)v_i \otimes c_i = \sum v_i \otimes c_i.$$

It follows $\sum(\pi(h)v_i - v_i) \otimes c_i = 0$, and therefore $\pi(h)v_i = v_i$, because c_i are linearly independent. Thus, $v_i \in V^H$, for all i, and hence $\mu \in V^H \otimes E$, proving the claim. From the claim, the lemma follows easily. $\square$

Proposition 6.16 *If F is algebraically closed and (π, V) is a smooth irreducible representation of G, then it is absolutely irreducible.*

Proof This is Proposition 2.2.6 in Casselman [17]. $\square$

Contragredient

Let (π, V) be a smooth representation of G. We denote by π^* the representation of G on the algebraic dual $V^* = \operatorname{Hom}_F(V, F)$ given by

$$(\pi^*(g)\ell)(v) = \ell(\pi(g^{-1})v),$$

for $g \in G$, $v \in V$, and $\ell \in V^*$. We denote by $\widetilde{V}$ the smooth part of V^* and by $\widetilde{\pi}$ the restriction of π^* to $\widetilde{V}$. We call $(\widetilde{\pi}, \widetilde{V})$ the **contragredient** of (π, V). The following are Propositions 2.1.10 and 2.1.11 from [17].

Proposition 6.17 *The following are equivalent*

(i) π is admissible;
(ii) $\widetilde{\pi}$ is admissible;
(iii) the contragredient of $\widetilde{\pi}$ is isomorphic to π, that is, $\widetilde{\widetilde{\pi}} \cong \pi$.

Proposition 6.18 *The functor $\pi \mapsto \tilde{\pi}$ is contravariant and exact.*

Tensor Product of Representations

Assume that we have another locally profinite group H. Then $G \times H$ is also locally profinite. If (π, U) is a representation of G and (τ, V) a representation of H, then $\pi \otimes \tau$ is the representation of $G \times H$ on the space $U \otimes V$ given by

$$(\pi \otimes \tau)(g, h)(u \otimes v) = \pi(g)u \otimes \tau(h)v$$

for $g \in G$, $h \in H$, $u \in U$, and $v \in V$.

Proposition 6.19

(i) *If (π_1, V_1) and (π_2, V_2) are irreducible admissible-smooth representations of G and H, respectively, then $(\pi_1 \otimes \pi_2, V_1 \otimes V_2)$ is an irreducible admissible-smooth representation of $G \times H$.*

(ii) *If (π, V) is an irreducible admissible-smooth representation of $G \times H$, then there exist irreducible admissible-smooth representations (π_1, V_1) and (π_2, V_2) of G and H, respectively, such that $\pi \cong \pi_1 \otimes \pi_2$.*

Proof (i) is Proposition 2.6.3 of [17] and (ii) is Proposition 2.6.4 of [17]. □

If η is a character of G and (π, V) a representation of G, then it is customary to denote by $\eta \otimes \pi$ the representation of G on V defined by

$$(\eta \otimes \pi)(g) = \eta(g)\pi(g), \quad g \in G.$$

6.2.4 Admissible-Smooth Representations

The following two results are well-known for complex representations; we reprove them here for representations over F.

Lemma 6.20 *Let H be a compact open subgroup of G. Denote by $\widehat{H}$ the set of equivalence classes of irreducible smooth representations of H. Then*

(i) *$\widehat{H}$ is countable;*
(ii) *every $\rho \in \widehat{H}$ is finite-dimensional.*

Proof For a compact open normal subgroup N of H, let us denote by $\widehat{H}_N$ the set of all $\rho \in \widehat{H}$ such that the restriction of ρ to N is trivial. If V is a representative of an equivalence class in $\widehat{H}_N$, then we can consider V as a representation of the finite group H/N. This representation is also irreducible, so it is finite-dimensional. By Maschke's theorem [24, Theorem 10.8], the group ring $F[H/N]$ is semisimple. Theorem 4.3 of [43, XVII, §4] implies that there is only a finite number of equivalence classes of simple $F[H/N]$-modules. Hence, $\widehat{H}_N$ is finite.

Define $H_n = H \cap G_n$. Then $\{H_n \mid n \in \mathbb{N}\}$ is a neighbourhood basis of identity in H consisting of compact normal subgroups. Every $\rho \in \widehat{H}$ is contained in $\widehat{H}_{H_n}$, for some n. On the other hand, each $\widehat{H}_{H_n}$ is finite. This implies that $\widehat{H}$ is countable. $\quad\square$

Proposition 6.21 *Let (π, V) be an admissible-smooth representation of G on an F-vector space V. Let H be a compact open subgroup of G. Then V is isomorphic to a countable direct sum*

$$V \cong \bigoplus_\rho m(\rho)\rho$$

where ρ runs over a set of representatives of $\widehat{H}$, and the multiplicity $m(\rho) < \infty$.

Proof Take an arbitrary nonzero vector $v \in V$. Then $v \in V^N$, for some compact open subgroup N of H. We may assume N is normal in H; if not, we can replace it by $\bigcap_{h \in H/N} hNh^{-1}$.

Denote by U_v the H-subrepresentation of V generated by v. We can consider U_v as a representation of the finite group H/N. Then U_v is finite-dimensional, because it is spanned over K by the finite set $\{\pi(\bar{h})v \mid \bar{h} \in H/N\}$. Maschke's theorem [24, Theorem 10.8] tells us that U_v is completely reducible as an H/N-representation. Then it is completely reducible as an H-representation as well.

Now, $V = \sum_{v \in V} U_v$ is a sum of irreducible H-representations. Theorem 15.3 of [24] implies that V is a direct sum of irreducible H-subrepresentations, so we can write $V = \bigoplus_{i \in I} U_i$. For an irreducible representation ρ of H, set

$$V(\rho) = \bigoplus_{U_i \cong \rho} U_i.$$

This is the ρ-isotypic component of V. There exists a compact open normal subgroup N of H such that $\rho|_N$ is trivial. Then $U_i \subseteq V^N$ for all $U_i \cong \rho$. It follows $V(\rho) \subseteq V^N$, and hence $\dim V(\rho) < \infty$. The set $\{i \in I \mid U_i \cong \rho\}$ is finite and $V(\rho) \cong m(\rho)\rho$, where $m(\rho) < \infty$. Finally,

$$V = \bigoplus_\rho V(\rho) \cong \bigoplus_\rho m(\rho)\rho.$$

The sum is countable because $\widehat{H}$ is countable. $\quad\square$

6.2.5 Smooth Principal Series

From now on, L is a finite extension of $\mathbb{Q}_p$ and $G = \mathbf{G}(L)$ is the group of L-points of a split connected reductive $\mathbb{Z}$-group $\mathbf{G}$.

Let $\chi : T \to F^\times$ be a smooth character of T. We extend it trivially to U and obtain a smooth character of $P = TU$, which we denote by the same letter χ. We denote by

$$\mathrm{Ind}_P^G(\chi)^{\mathrm{sm}}$$

the space of all functions $f : G \to F$ satisfying

(I) $f(gp) = \chi(p)^{-1} f(g)$ for all $p \in P$, $g \in G$, and
(II) there exists a compact open subgroup H_f of G such that $f(hg) = f(g)$, for all $h \in H_f$, $g \in G$.

The group G acts on $\mathrm{Ind}_P^G(\chi)^{\mathrm{sm}}$ by left translation $(L_g f) = f(g^{-1}x)$. Notice that condition (II) can be written as $L_h f = f$, for all $h \in H_f$. This means that f belongs to the space of H_f-fixed vectors, and hence $\mathrm{Ind}_P^G(\chi)^{\mathrm{sm}}$ is a smooth representation of G. We call it the **smooth principal series** induced by χ from P to G.

There is a more general concept of parabolically induced representations, where the minimal parabolic subgroup P is replaced by an arbitrary parabolic subgroup Q, and the character χ is replaced by a smooth representation of the Levi factor of Q. For details, see Casselman [17, Section 3].

Remark 6.22 The definition of $\mathrm{Ind}_P^G(\chi)^{\mathrm{sm}}$ is with respect to the action of G by left translation. If we instead want to use the right translation, we proceed as follows. Let V be the space of functions $f : G \to F$ satisfying

(I') $f(pg) = \chi(p)f(g)$ for all $p \in P$, $g \in G$, and
(II') there exists a compact open subgroup H_f of G such that $f(gh) = f(g)$, for all $h \in H_f$, $g \in G$.

Then G acts on V by $(R_g f)(x) = f(xg)$. Similarly, the algebraic induction could be defined using the right translation.

To explain how the two forms of induction relate, let $U = \mathrm{Ind}_P^G(\chi)^{\mathrm{sm}}$. For $f \in U$, define $\varphi(f) : G \to F$ by $\varphi(f)(g) = f(g^{-1})$. Then (II) for f implies (II') for $\varphi(f)$. Also,

$$\varphi(f)(pg) = f(g^{-1}p^{-1}) = \chi(p)f(g^{-1}) = \chi(p)\varphi(f)(g),$$

for any $p \in P$ and $g \in G$, proving (I'). This shows $\varphi(f) \in V$. The corresponding map $\varphi : U \to V$ is clearly a linear bijection. It is easy to check that $\varphi \circ L_g = R_g \circ \varphi$. Hence, φ is an intertwining operator, and the representations U and V are equivalent.

Although the character χ is trivial on U, the induced representation $\mathrm{Ind}_P^G(\chi)^{\mathrm{sm}}$ depends on the choice of $P = TU \supset T$. We start with the following observation.

Exercise 6.23 Suppose $\sigma : G \to G$ is an automorphism of G. If π is a representation of G, we denote by $\sigma\pi$ the representation of G defined by $\sigma\pi(g) =$

$\pi(\sigma^{-1}(g))$. Let $\chi : T \to F^\times$ be a smooth character of T. Prove that

$$\mathrm{Ind}_{\sigma(P)}^G(\sigma\chi)^{\mathrm{sm}} \cong \sigma\,\mathrm{Ind}_P^G(\chi)^{\mathrm{sm}}.$$

If we apply the above exercise to the conjugation by $w \in W$, we get

$$\mathrm{Ind}_{wPw^{-1}}^G(w\chi)^{\mathrm{sm}} \cong \mathrm{Ind}_P^G(\chi)^{\mathrm{sm}},$$

where $w\chi$ denotes the character of T defined by

$$w\chi(t) = \chi(w^{-1}tw), \quad t \in T.$$

Then, if we want to compare the induction from P with the induction from the opposite Borel subgroup P^-, we have to conjugate by the longest element $w_\ell \in W$.

Proposition 6.24 *Let* $\chi : T \to F^\times$ *be a smooth character of T and* $\eta : G \to F^\times$ *a smooth character of G. Then*

$$\mathrm{Ind}_P^G(\eta \otimes \chi)^{\mathrm{sm}} = \eta \otimes \mathrm{Ind}_P^G(\chi)^{\mathrm{sm}}.$$

Proof Notice that $\eta \otimes \chi = \eta\chi$. Denote by U the space of $\mathrm{Ind}_P^G(\chi)^{\mathrm{sm}}$ and by V the space of $\mathrm{Ind}_P^G(\eta\chi)^{\mathrm{sm}}$. Then $\pi = \eta \otimes \mathrm{Ind}_P^G(\chi)^{\mathrm{sm}}$ is the representation of G on U given by $\pi(g) = \eta(g)L_g$, or

$$\pi(g)f(x) = \eta(g)f(g^{-1}x)$$

for $g, x \in G$ and $f \in U$. For $f \in U$, define $\varphi(f) : G \to F$ by

$$\varphi(f)(x) = \eta(x^{-1})f(x).$$

Then $\varphi(f)(gp) = \eta((gp)^{-1})f(gp) = (\eta\chi)(p^{-1})\varphi(f)(g)$. It follows $\varphi(f) \in V$, and $f \mapsto \varphi(f)$ defines a map $\varphi : U \to V$. The map is clearly linear and bijective.

We claim $\varphi \in \mathrm{Hom}_G(\pi, V)$. To prove it, we have to show that $\varphi(\pi(g)f) = L_g\varphi(f)$ for all $f \in U$, $g \in G$. Evaluating the left hand side at $x \in G$, we get

$$\varphi(\pi(g)f)(x) = \eta(x^{-1})\pi(g)f(x) = \eta(x^{-1})\eta(g)f(g^{-1}x).$$

On the other hand,

$$(L_g\varphi(f))(x) = \varphi(f)(g^{-1}x) = \eta(x^{-1}g)f(g^{-1}x).$$

This proves the claim, thus proving that the representations π and V are equivalent. $\qquad\square$

Exercise 6.25 Let $\chi : T \to F^\times$ be a smooth character of T. If E/F is a field extension, prove that

$$\mathrm{Ind}_P^G(\chi \otimes E)^{\mathrm{sm}} = \mathrm{Ind}_P^G(\chi)^{\mathrm{sm}} \otimes E.$$

Notice that $\chi \otimes E$ is simply χ taken as $\chi : T \to E^\times$. Using alternate notation, the above equality is written as $\mathrm{Ind}_P^G(\chi_E)^{\mathrm{sm}} = \mathrm{Ind}_P^G(\chi)_E^{\mathrm{sm}}$.

Let us compare the algebraic induction $\mathrm{ind}_{\mathbf{P}}^{\mathbf{G}}(\mu)$ to $\mathrm{Ind}_P^G(\chi)^{\mathrm{sm}}$. Assume $L \subseteq F$, because otherwise, there are no algebraic representations of $G = \mathbf{G}(L)$ on F-spaces except constants.

We write $\mathrm{ind}_P^G(\mu)$ for $\mathrm{ind}_{\mathbf{P}}^{\mathbf{G}}(\mu)(F)$. The algebraic induction starts with an algebraic character μ, while the smooth induction starts with a smooth character χ. The only character of T which is both algebraic and locally constant is the trivial character $\mathbf{1}$. Recall that the group $X(\mathbf{T})$ of algebraic characters of $\mathbf{T}$ is written additively. Then the trivial character $\mathbf{1}$ corresponds to $0 \in X(\mathbf{T})$. The simple module $V = L(0)$ has the highest weight 0. From Lemma 6.9, we see that 0 is the only weight of V. Then $V = V_0$, the weight space corresponding to the weight 0. It follows $L(0) = \mathbf{1}$ and

$$\mathrm{ind}_P^G(\mathbf{1}) = \mathbf{1},$$

where $\mathbf{1}$ on the right is the trivial one-dimensional representation of G. On the other hand, $\mathrm{Ind}_P^G(\mathbf{1})^{\mathrm{sm}}$ consists of smooth functions. Only constant functions are both smooth and algebraic. We have

$$\mathrm{ind}_P^G(\mathbf{1}) \hookrightarrow \mathrm{Ind}_P^G(\mathbf{1})^{\mathrm{sm}}.$$

Normalized Induction

Normalization is very important for complex smooth representations because normalized induction preserves complex unitarity (see [17, Proposition 3.1.4]). Here, we introduce normalization so that certain statements have nice and symmetric form (see Propositions 6.26, 6.28, 6.31, and 6.33). For normalization, we have to assume that F contains a square root of q_L, which we fix and denote by $q_L^{1/2}$.

Let $\mathfrak{u} = \mathrm{Lie}(U)$ be the Lie algebra of the unipotent radical U of P. Then P acts on $\mathfrak{u}$ by adjoint action. For $p \in P$, define

$$\delta_P(p) = |\det \mathrm{Ad}_{\mathfrak{u}}(p)|_L$$

as in [17, Section 1.5] or [69, (2.3.6)]. Then $\delta_P : P \to F^\times$ is a smooth character of P called the **modulus character**. (See Exercise 6.30 for an explicit description of δ_P for $GL_2(L)$.) It is trivial on U, hence essentially a character of T. If χ is a smooth character of T, we define

$$i_{G,P}(\chi) = \text{Ind}_P^G(\delta_P^{1/2}\chi)^{\text{sm}}.$$

This is called **normalized induction**. Some authors define the modulus character as the inverse of our δ_P and put $\delta_P^{-1/2}$ in the definition of normalized induction, thus getting the same $i_{G,P}(\chi)$ (see [53, Remarque II.3.7]).

Proposition 6.26 *The contragredient of $i_{G,P}(\chi)$ is isomorphic to $i_{G,P}(\chi^{-1})$.*

Proof Follows from Theorem 2.4.2 of [17], since the contragredient of χ is equal to χ^{-1}. □

Composition Factors of Principal Series

Let (π, V) be a smooth representation of G. Suppose there exists a finite sequence

$$0 = V_0 \subset V_1 \subset \cdots \subset V_\ell = V \tag{6.5}$$

of G-invariant subspaces of V such that V_i/V_{i-1} is irreducible for $i = 1, \ldots, \ell$. The sequence (6.5) is called a **composition series** or a **Jordan-Hölder series** of V.

Exercise 6.27 Suppose $0 = V_0 \subset V_1 \subset \cdots \subset V_\ell = V$ and $0 = U_0 \subset U_1 \subset \cdots \subset U_m = V$ are two composition series of a smooth representation (π, V) of G. Prove that $\ell = m$ and that there is some permutation σ of $\{1, \ldots, \ell\}$ such that

$$U_{\sigma(i)}/U_{\sigma(i)-1} \cong V_i/V_{i-1}, \quad 1 \le i \le \ell.$$

Based on Exercise 6.27, if (π, V) is a smooth representation of G with composition series (6.5), we say that ℓ is the **length** of V, and V_i/V_{i-1} are the **composition factors**.

Proposition 6.28 *Assume $F = \mathbb{C}$ or $\overline{\mathbb{Q}}_p$. Let χ be a smooth character of T. Then,*

(i) The length of $i_{G,P}(\chi)$ is at most the order of the Weyl group.
(ii) Let $w \in W$. Then, the composition factors of $i_{G,P}(\chi)$ and $i_{G,P}(w\chi)$ are same.
(iii) If π is a composition factor of $i_{G,P}(\chi)$, then there exists $w \in W$ and an embedding of π into $i_{G,P}(w\chi)$.

Proof For $F = \mathbb{C}$, the assertions follow from [17], Corollary 6.3.9 and Theorem 6.3.11. The result for $F = \overline{\mathbb{Q}}_p$ then follows from the proof of Proposition 6.14. Namely, if we fix an isomorphism of abstract fields $\iota : \mathbb{C} \to \overline{\mathbb{Q}}_p$, then every representation $(\pi, V) \in \text{Rep}_{\overline{\mathbb{Q}}_p}^{\text{sm}}(G)$ can be considered as a representation in $\text{Rep}_{\mathbb{C}}^{\text{sm}}(G)$. □

Corollary 6.29 *Let $F = K$, a finite extension of $\mathbb{Q}_p$. Let $\chi : T \to K^\times$ be a smooth character. Then the length of $i_{G,P}(\chi)$ is at most the order of the Weyl group.*

Proof The length of $i_{G,P}(\chi)$ is less than or equal to the length of $i_{G,P}(\chi) \otimes_F \overline{\mathbb{Q}}_p$. The latter representation is isomorphic to $i_{G,P}(\chi \otimes_F \overline{\mathbb{Q}}_p)$ by Exercise 6.25, and its length is at most the order of the Weyl group by Proposition 6.28. □

6.2.6 Smooth Principal Series of $GL_2(L)$ and $SL_2(L)$

In this section, $G = GL_2(L)$ and $H = SL_2(L)$. We consider representations on K-vector spaces, where K is a finite extension of $\mathbb{Q}_p$. We assume that K contains a square root of q_L, which we fix and denote by $q_L^{1/2}$. Denote by T the diagonal torus in G and by $P \subset G$ the group of upper triangular matrices.

Exercise 6.30 We defined the modulus character on P as $\delta_P(p) = |\det \mathrm{Ad}_u(p)|_L$. Prove that it is given by

$$\delta_P\left(\begin{pmatrix} a & b \\ 0 & c \end{pmatrix}\right) = \left|\frac{a}{c}\right|_L.$$

Given a smooth character $\chi : T \to K^\times$, we consider the normalized principal series $i_{G,P}(\chi)$. The groups G and H have the same Weyl group. It is isomorphic to the symmetric group S_2, so it has two elements. Corollary 6.29 then implies that the length of a principal series representation of G or H is at most two.

Write $W = \{1, w\}$. A smooth character $\chi : T \to K^\times$ can be written as $\chi = \chi_1 \otimes \chi_2$, where χ_1 and χ_2 are characters of $L^\times$. Then $w\chi = \chi_2 \otimes \chi_1$.

We first consider the special case when $\chi = \delta_P^{-\frac{1}{2}}$. Then

$$i_{G,P}(\delta_P^{-\frac{1}{2}}) = \mathrm{Ind}_P^G(\mathbf{1})^{\mathrm{sm}}.$$

As explained earlier, this representation contains the trivial one-dimensional representation $\mathbf{1}$. By the above discussion, the length of $i_{G,P}(\delta_P^{-\frac{1}{2}})$ is two. It follows that the quotient $i_{G,P}(\delta_P^{-\frac{1}{2}})/\mathbf{1}$ is irreducible. It is called the **Steinberg representation** and it is denoted by St. We have the following exact sequence

$$0 \to \mathbf{1} \to i_{G,P}(\delta_P^{-\frac{1}{2}}) \to \mathrm{St} \to 0. \tag{6.6}$$

The sequence does not split over $\overline{\mathbb{Q}}_p$ by Zelevinsky [80, Proposition 1.11 (b)] and consequently, it does not split over K. Furthermore, taking contragredient and using Propositions 6.18 and 6.26, we obtain

$$0 \to \widetilde{\mathrm{St}} \to i_{G,P}(\delta_P^{\frac{1}{2}}) \to \mathbf{1} \to 0. \tag{6.7}$$

Notice that $w(\delta_P^{-\frac{1}{2}}) = \delta_P^{\frac{1}{2}}$. By Proposition 6.28, the composition factors of $i_{G,P}(\delta_P^{-\frac{1}{2}})_{\overline{\mathbb{Q}}_p}$ and $i_{G,P}(\delta_P^{\frac{1}{2}})_{\overline{\mathbb{Q}}_p}$ are isomorphic. Consequently, $\widetilde{\mathrm{St}}_{\overline{\mathbb{Q}}_p} \cong \mathrm{St}_{\overline{\mathbb{Q}}_p}$. Then Lemma 5.1 of [52] tells us that $\widetilde{\mathrm{St}} \cong \mathrm{St}$.

Proposition 6.31 *Let $\chi_1 \otimes \chi_2 : T \to K^\times$ be a smooth character. The representation $i_{G,P}(\chi_1 \otimes \chi_2)$ is reducible if and only if $\chi_1\chi_2^{-1} = |\ |_L^{\pm 1}$.*

(i) *If $\chi_1\chi_2^{-1} = |\ |_L^{-1}$, then there is a smooth character η such that $(\chi_1, \chi_2) = (\eta|\ |_L^{-\frac{1}{2}}, \eta|\ |_L^{\frac{1}{2}})$. The representation $i_{G,P}(\eta|\ |_L^{-\frac{1}{2}} \otimes \eta|\ |_L^{\frac{1}{2}})$ fits in the following exact sequence*

$$0 \to \eta \circ \det \to i_{G,P}(\eta|\ |_L^{-\frac{1}{2}} \otimes \eta|\ |_L^{\frac{1}{2}}) \to (\eta \circ \det) \otimes \mathrm{St} \to 0.$$

The above sequence does not split.

(ii) *If $\chi_1\chi_2^{-1} = |\ |_L$, then there is a smooth character η such that $(\chi_1, \chi_2) = (\eta|\ |_L^{\frac{1}{2}}, \eta|\ |_L^{-\frac{1}{2}})$. The representation $i_{G,P}(\eta|\ |_L^{\frac{1}{2}} \otimes \eta|\ |_L^{-\frac{1}{2}})$ fits in the following exact sequence*

$$0 \to (\eta \circ \det) \otimes \mathrm{St} \to i_{G,P}(\eta|\ |_L^{\frac{1}{2}} \otimes \eta|\ |_L^{-\frac{1}{2}}) \to \eta \circ \det \to 0.$$

The above sequence does not split.

Proof If we consider representations over $\overline{\mathbb{Q}}_p$, then we know from [80, Proposition 1.11 (a)] that $i_{G,P}(\chi_1 \otimes \chi_2)_{\overline{\mathbb{Q}}_p}$ is reducible if and only if $\chi_1\chi_2^{-1} = |\ |_L^{\pm 1}$. Consequently, over K, if $\chi_1\chi_2^{-1} \neq |\ |_L^{\pm 1}$, then $i_{G,P}(\chi_1 \otimes \chi_2)$ is irreducible.

Assume $(\chi_1, \chi_2) = (\eta|\ |_L^{-\frac{1}{2}}, \eta|\ |_L^{\frac{1}{2}})$. We can write

$$\chi_1 \otimes \chi_2 = (\eta \circ \det) \otimes \delta_P^{-\frac{1}{2}}.$$

If we twist the exact sequence (6.6) by $\eta \circ \det$ and apply Proposition 6.24, we obtain assertion (i). Similarly, assertion (ii) follows from the exact sequence (6.7). $\qquad\square$

For future reference, we state the same result using unnormalized induction. The statement can also be found in Section 9.6 of [14]).

Corollary 6.32 *Let $\chi_1 \otimes \chi_2 : T \to K^\times$ be a smooth character. The representation $\mathrm{Ind}_P^G(\chi_1 \otimes \chi_2)^{\mathrm{sm}}$ is reducible if and only if $\chi_1 = \chi_2$ or $\chi_1\chi_2^{-1} = |\ |_L^2$. In particular, $\mathrm{Ind}_P^G(1)^{\mathrm{sm}}$ and $\mathrm{Ind}_P^G(\delta_P)^{\mathrm{sm}} = \mathrm{Ind}_P^G(|\ |_L \otimes |\ |_L^{-1})^{\mathrm{sm}}$ are reducible.*

Next, we consider $H = SL_2(L)$. Let $P_H = P \cap H$. Let $T_H = T \cap H$ be the torus of diagonal matrices in H, which is isomorphic to $L^\times$ via the map $a \mapsto \mathrm{diag}(a, a^{-1})$. Given a character χ of T_H, we now consider $i_{H,P_H}(\chi)$. Since

$\delta^{\frac{1}{2}}(\mathrm{diag}(a, a^{-1})) = |a|_L$, we see that $i_{H,P_H}(\chi)$ does not depend on the choice of the square root of q that we fixed in the beginning.

Restricting the exact sequence (6.6) to H, we get

$$0 \to \mathbf{1} \to i_{H,P_H}(\mid\mid_L^{-1}) \to \mathrm{St} \to 0.$$

Similarly, restricting (6.7) to H, we get

$$0 \to \mathrm{St} \to i_{H,P_H}(\mid\mid_L) \to \mathbf{1} \to 0.$$

Proposition 6.33 *Let* $\chi : T_H \to K^\times$ *be a smooth character. Suppose that the smooth normalized induced representation* $i_{H,P_H}(\chi)$ *is reducible. Then either* $\chi = \mid\mid_L^{\pm 1}$ *or* $\chi^2 = \mathbf{1}$, $\chi \neq \mathbf{1}$.

Proof For representations over $\overline{\mathbb{Q}}_p$, we know that $i_{H,P_H}(\chi)_{\overline{\mathbb{Q}}_p}$ is reducible if and only if either $\chi = \mid\mid_L^{\pm 1}$ or $\chi^2 = \mathbf{1}$, $\chi \neq \mathbf{1}$. This follows from the discussion in [34, ch. 2, §3–5]. It is assumed in [34] that the residue characteristic is odd, but the properties of these representations (in particular, their reducibility) also hold for residue characteristic two (see also [34, ch. 2, §8]). Irreducibility over $\overline{\mathbb{Q}}_p$ implies irreducibility over K, proving the proposition. $\square$

Suppose that χ is a nontrivial quadratic character. Then $i_{H,P_H}(\chi)_{\overline{\mathbb{Q}}_p}$ is a direct sum of two components, and these components are inequivalent by Tadić [74, 1.2]. It is remarked in Section 4 of [68] that $i_{H,P_H}(\chi)$ over K could be irreducible (but, of course, not absolutely irreducible).

Chapter 7
Continuous Principal Series

In this chapter, we use the notation listed in *Notation in Part II* (see page 89). In particular, $\mathbb{Q}_p \subseteq L \subseteq K$ is a sequence of finite extensions, $\mathbf{G}$ is a split connected reductive $\mathbb{Z}$-group and $\mathbf{P} = \mathbf{TU}$ is a Borel subgroup of $\mathbf{G}$.

We study the continuous principal series of $G = \mathbf{G}(L)$ and $G_0 = \mathbf{G}(o_L)$. Our approach is based on the Schneider-Teitelbaum duality from Chap. 4. More specifically, we use Theorem 4.43, which tells us that the duality map $V \mapsto V'$ is an anti-equivalence between the category of admissible Banach space representations of G_0 and the category of finitely generated $K[[G_0]]$-Iwasawa modules.

Let $\chi : P \to K^\times$ be a continuous character and χ_0 its restriction to $P_0 = \mathbf{P}(o_L)$. In Sect. 7.1, we establish some basic properties of the continuous principal series $\mathrm{Ind}_{P_0}^{G_0}(\chi_0^{-1})$ and $\mathrm{Ind}_P^G(\chi^{-1})$. In particular, we prove that they are Banach. After that, we work on the dual side and study the corresponding $(K[[G_0]], G)$-modules. In Theorem 7.12, we prove that the dual of $\mathrm{Ind}_{P_0}^{G_0}(\chi^{-1})$ is isomorphic to

$$M^{(\chi)} = K[[G_0]] \otimes_{K[[P_0]]} K^{(\chi)}.$$

In the rest of the chapter, we describe the structure of $M^{(\chi)}$. In Proposition 7.14, we introduce a decomposition

$$M^{(\chi)} = \bigoplus_{w \in W} M_w^{(\chi)}$$

into components $M_w^{(\chi)}$ indexed by the Weyl group W of $\mathbf{G}$. In Proposition 7.20, we give a projective limit realization of $M_0^{(\chi)} = o_K[[G_0]] \otimes_{o_K[[P_0]]} o_K^{(\chi)}$, namely

$$M_0^{(\chi)} \cong \varprojlim_{n \in \mathbb{N}} \left(o_K[G_0/G_n] \otimes_{o_K[P_0]} o_K^{(\chi)} \right).$$

© The Author(s), under exclusive license to Springer Nature Switzerland AG 2022
D. Ban, *p-adic Banach Space Representations*, Lecture Notes
in Mathematics 2325, https://doi.org/10.1007/978-3-031-22684-7_7

In Sect. 7.4, we describe $M_w^{(\chi)}$ as a tensor product, thus obtaining a $K[[B]]$-module decomposition

$$M^{(\chi)} \cong \bigoplus_{w \in W} K[[B]] \otimes_{K[[P_{\frac{1}{2}}^{w,\pm}]]} K^{(w\chi)}$$

7.1 Continuous Principal Series Are Banach

Let $\chi : T_0 \to K^\times$ be a continuous character of T_0. Since T_0 is compact, the image of χ is compact, and hence it must be contained in $o_K^\times$. We consider the **continuous principal series representation**

$$\mathrm{Ind}_{P_0}^{G_0}(\chi^{-1}) = \{f \in C(G_0, K) \mid f(gp) = \chi(p)f(g) \text{ for all } p \in P_0, g \in G_0\},$$

with the action of G_0 by left translation $(L_g f)(x) = f(g^{-1}x)$. The space $\mathrm{Ind}_{P_0}^{G_0}(\chi^{-1})$ is a closed subspace of the Banach space $C(G_0, K)$, so it is itself a Banach space. As introduced in Sect. 3.2, the Banach topology on $C(G_0, K)$ is induced by the supremum norm

$$\|f\| = \sup_{g \in G_0} |f(g)|.$$

By Exercise 4.14, the action of G_0 on $C(G_0, K)$ by left and right translations is continuous. It follows that the action of G_0 on $\mathrm{Ind}_{P_0}^{G_0}(\chi^{-1})$ is also continuous.

Similarly, we define the continuous principal series of G. If $\chi : T \to K^\times$ is a continuous character, we define

$$\mathrm{Ind}_P^G(\chi^{-1}) = \{f : G \to K \text{ continuous} \mid f(gp) = \chi(p)f(g) \text{ for all } p \in P, g \in G\}$$

with the action of G by left translations.

While the Banach space structure on $\mathrm{Ind}_{P_0}^{G_0}(\chi^{-1})$ comes from the embedding $\mathrm{Ind}_{P_0}^{G_0}(\chi^{-1}) \subset C(G_0, K)$, for $\mathrm{Ind}_P^G(\chi^{-1})$, we do not have a natural ambient Banach space. Instead, we proceed as follows. Set $\chi_0 = \chi|_{T_0}$. If $f \in \mathrm{Ind}_P^G(\chi^{-1})$, then clearly $f|_{G_0} \in \mathrm{Ind}_{P_0}^{G_0}(\chi_0^{-1})$. We will prove that the map $f \mapsto f|_{G_0}$ is a bijection from $\mathrm{Ind}_P^G(\chi^{-1})$ to $\mathrm{Ind}_{P_0}^{G_0}(\chi_0^{-1})$ and we will use this bijection to define a Banach space structure on $\mathrm{Ind}_P^G(\chi^{-1})$ (Proposition 7.3). The proof that this structure, with the action of G by left translations, is a Banach space representation of G is given in Proposition 7.5.

7.1.1 Direct Sum Decomposition of $\mathrm{Ind}_{P_0}^{G_0}(\chi_0^{-1})$

Exercise 7.1 Suppose $G_0 = \coprod_{i=1}^{r} S_i$ is a disjoint union of compact open subsets. Prove that $f \mapsto \sum_{i=1}^{r} f|_{S_i}$ defines an isomorphism of Banach spaces $C(G_0, K) \cong \bigoplus_{i=1}^{r} C(S_i, K)$, where the topology of the direct sum comes from the norm defined by Eq. (A.3) in Appendix A.3.

Let $\chi : T \to K^\times$ be a continuous character of T and $\chi_0 = \chi|_{T_0}$. Recall the disjoint union

$$G_0 = \coprod_{w \in W} \dot{w} U^-_{w,\frac{1}{2}} P_0$$

as in (5.7). If we apply Exercise 7.1 to the above disjoint union, we see that $\mathrm{Ind}_{P_0}^{G_0}(\chi_0^{-1})$ decomposes as

$$\mathrm{Ind}_{P_0}^{G_0}(\chi_0^{-1}) = \bigoplus_{w \in W} \mathrm{Ind}_{P_0}^{G_0}(\chi_0^{-1})_w, \tag{7.1}$$

where

$$\mathrm{Ind}_{P_0}^{G_0}(\chi_0^{-1})_w = \{f \in \mathrm{Ind}_{P_0}^{G_0}(\chi_0^{-1}) \mid \mathrm{supp}(f) \subset w U^-_{w,\frac{1}{2}} P_0\}.$$

Proposition 7.2

(i) *For every $w \in W$, we have the following isomorphism of topological vector spaces:*

$$\mathrm{Ind}_{P_0}^{G_0}(\chi_0^{-1})_w \cong C(U^-_{w,\frac{1}{2}}, K).$$

(ii)

$$\mathrm{Ind}_{P_0}^{G_0}(\chi_0^{-1}) = \bigoplus_{w \in W} \mathrm{Ind}_{P_0}^{G_0}(\chi_0^{-1})_w \cong \bigoplus_{w \in W} C(U^-_{w,\frac{1}{2}}, K).$$

Proof

(i) Recall that we fixed a set $\dot{W}$ of representatives for the elements of W. Let $\dot{w}$ be the representative for w. Notice that an element of $\mathrm{Ind}_{P_0}^{G_0}(\chi_0^{-1})_w$ is determined by its restriction to $\dot{w} U^-_{w,\frac{1}{2}}$, which is continuous. Moreover, given a continuous function $h : U^-_{w,\frac{1}{2}} \to K$, we may define

$$
f_h(g) = \begin{cases} h(u)\chi_0(p), & g = \dot{w}up, \ u \in U^-_{w,\frac{1}{2}}, \ p \in P_0 \\ 0, & g \notin wU^-_{w,\frac{1}{2}}P_0. \end{cases} \tag{7.2}
$$

Then $h \mapsto f_h$ is a K-linear bijection $C(U^-_{w,\frac{1}{2}}, K) \to \operatorname{Ind}^{G_0}_{P_0}(\chi_0^{-1})_w$. Since $\chi(P_0) \subseteq o_K^\times$, it follows $\|f_h\| = \|h\|$, and hence $h \mapsto f_h$ is an isomorphism of topological vector spaces.

(ii) Follows from (i) and Eq. (7.1).

□

From Proposition 5.45, we know that G/P and G_0/P_0 have the same set of coset representatives. The disjoint union $G = \coprod_{w \in W} \dot{w} U^-_{w,\frac{1}{2}} P$ implies that $\operatorname{Ind}^G_P(\chi^{-1})$ decomposes as the direct sum

$$
\operatorname{Ind}^G_P(\chi^{-1}) = \bigoplus_{w \in W} \operatorname{Ind}^G_P(\chi^{-1})_w, \tag{7.3}
$$

where

$$
\operatorname{Ind}^G_P(\chi^{-1})_w = \{f \in \operatorname{Ind}^G_P(\chi^{-1}) \mid \operatorname{supp}(f) \subset wU^-_{w,\frac{1}{2}}P\}.
$$

So far, we have not defined the topology on $\operatorname{Ind}^G_P(\chi^{-1})$. From the above decomposition, it is natural to equip $\operatorname{Ind}^G_P(\chi^{-1})$ with the topology coming from identifying $\operatorname{Ind}^G_P(\chi^{-1})_w$ with $C(U^-_{w,\frac{1}{2}}, K)$.

Proposition 7.3 *Let $\chi : T \to K^\times$ be a continuous character.*

(i) $\operatorname{Ind}^G_P(\chi^{-1})$ *is a Banach space with norm*

$$
\|f\| = \sup_{g \in G_0} |f(g)|. \tag{7.4}
$$

(ii) *The restriction $f \mapsto f|_{G_0}$ is a topological isomorphism from $\operatorname{Ind}^G_P(\chi^{-1})$ to $\operatorname{Ind}^{G_0}_{P_0}(\chi_0^{-1})$.*

(iii)

$$
\operatorname{Ind}^G_P(\chi^{-1}) = \bigoplus_{w \in W} \operatorname{Ind}^G_P(\chi^{-1})_w \cong \bigoplus_{w \in W} C(U^-_{w,\frac{1}{2}}, K).
$$

Proof Similarly as in the proof of Proposition 7.2, we can define a bijection $h \mapsto F_h$ from $C(U^-_{w,\frac{1}{2}}, K)$ to $\mathrm{Ind}^G_P(\chi^{-1})_w$. We use this bijection to define the topology on $\mathrm{Ind}^G_P(\chi^{-1})_w$. We will prove that it is equivalent to the Banach topology induced by the norm (7.4). We have

$$\mathrm{Ind}^G_P(\chi^{-1})_w \cong C(U^-_{w,\frac{1}{2}}, K) \cong \mathrm{Ind}^{G_0}_{P_0}(\chi_0^{-1})_w,$$

where the second isomorphism follows from Proposition 7.2 (i). Moreover, given $h \in C(U^-_{w,\frac{1}{2}}, K)$, the corresponding functions $F_h \in \mathrm{Ind}^G_P(\chi^{-1})_w$ and $f_h \in \mathrm{Ind}^{G_0}_{P_0}(\chi_0^{-1})_w$ satisfy $F_h|_{G_0} = f_h$. It follows that the restriction $f \mapsto f|_{G_0}$ is a topological isomorphism from $\mathrm{Ind}^G_P(\chi^{-1})_w$ to $\mathrm{Ind}^{G_0}_{P_0}(\chi_0^{-1})_w$. Then Eqs. (7.1) and (7.3) imply

$$\mathrm{Ind}^G_P(\chi^{-1}) \cong \mathrm{Ind}^{G_0}_{P_0}(\chi_0^{-1}),$$

with the isomorphism given by the restriction $f \mapsto f|_{G_0}$. Then the supremum norm on $\mathrm{Ind}^{G_0}_{P_0}(\chi_0^{-1})$ induces the norm (7.4) on $\mathrm{Ind}^G_P(\chi^{-1})$. $\qquad\square$

Parts (i) and (ii) in Proposition 7.3 still hold if we replace G_0 by an arbitrary compact subgroup H of G satisfying the Iwasawa decomposition $G = HP$. We define the corresponding norm on $\mathrm{Ind}^G_P(\chi^{-1})$ by

$$\|f\|_H = \sup_{h \in H} |f(h)|.$$

If $H = G_0$, this norm is equal to the norm $\|\ \|$ introduced earlier. In the proof of the lemma below, we follow the approach from an unpublished note by Peter Schneider.

Lemma 7.4 *Let H be an arbitrary compact subgroup of G satisfying the Iwasawa decomposition $G = HP$.*

(i) *The norm $\|\ \|_H$ on $\mathrm{Ind}^G_P(\chi^{-1})$ is equivalent to $\|\ \|$.*
(ii) *The restriction $f \mapsto f|_H$ is a topological isomorphism from $\mathrm{Ind}^G_P(\chi^{-1})$ to $\mathrm{Ind}^H_{H\cap P}(\chi^{-1})$.*

Proof

(i) Since H is totally disconnected, the projection $H \to H/P$ has a continuous section. It follows that there exists a compact open subset $A \subset H$ such that the multiplication map induces a homeomorphism $A \times H \cap P \xrightarrow{\sim} H$. The Iwasawa decomposition $G = HP$ then induces a homeomorphism

$$A \times P \xrightarrow{\sim} G. \tag{7.5}$$

Let H' be another maximal compact subgroup of G such that $G = H'P$. We will prove that the norms $\|\ \|_H$ and $\|\ \|_{H'}$ are equivalent. By (7.5), for any $h' \in$

$\cdot H'$, there exist unique $h'_P \in P$ and $h'_A \in A$ such that $h' = h'_A h'_P$. Define

$$c = \sup_{h' \in H'} |\chi(h'_P)|.$$

Then $c \neq 0$ and moreover $c < \infty$, by compactness of H'. Take $f \in \mathrm{Ind}_P^G(\chi^{-1})$. Then for any $h' \in H'$, $|f(h')| = |f(h'_A h'_P)| = |\chi(h'_P)| \cdot |f(h'_A)| \leq c \cdot |f(h'_A)|$. Since $A \subset H$, it follows

$$\|f\|_{H'} = \sup_{h' \in H'} |f(h')| \leq c \cdot \sup_{h \in H} |f(h)| = c \cdot \|f\|_H.$$

Switching H and H', we see that there exists $c' > 0$ such that $\|f\|_H \leq c' \cdot \|f\|_{H'}$ for all $f \in \mathrm{Ind}_P^G(\chi^{-1})$. This proves that the norms $\|\ \|_H$ and $\|\ \|_{H'}$ are equivalent, also implying assertion (i).

(ii) By (i), we may consider $\mathrm{Ind}_P^G(\chi^{-1})$ equipped with the norm $\|\ \|_H$. Define

$$\varphi : \mathrm{Ind}_P^G(\chi^{-1}) \to C(A, K)$$

by $f \mapsto f|A$. Then φ is a linear map. Recall that $f \in \mathrm{Ind}_P^G(\chi^{-1})$ satisfies $f(gp) = \chi(p)f(g)$. This together with the homeomorphism $A \times P \xrightarrow{\sim} G$ implies that φ is bijective. As usual, we consider $C(A, K)$ equipped with the supremum norm. Notice that $\chi|_{H \cap P} : H \cap P \to K^\times$ has compact image, so the image must be contained in $o_K^\times$. It follows $|\chi(p)| = 1$ for all $p \in H \cap P$. Consequently, for $f \in \mathrm{Ind}_P^G(\chi^{-1})$ we have

$$\|f\|_H = \sup_{h \in H} |f(h)| = \sup_{(a,p) \in A \times (P \cap H)} |f(ap)| = \sup_{a \in A} |f(a)| = \|\varphi(f)\|.$$

It follows that φ is an isometry, and hence it is a topological isomorphism. We have

$$\mathrm{Ind}_P^G(\chi^{-1}) \cong C(A, K) \quad \text{and} \quad \mathrm{Ind}_{H \cap P}^H(\chi^{-1}) \cong C(A, K)$$

where both isomorphisms are given by restrictions and fit in the following commutative diagram of restriction maps

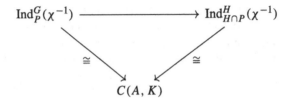

Then the top arrow is also an isomorphism, proving (ii). $\qquad\square$

Proposition 7.5 *The space* $\mathrm{Ind}_P^G(\chi^{-1})$, *with norm* (7.4) *and the G-action by left translations, is a K-Banach space representation of G.*

Proof We know from Proposition 7.3 that $\mathrm{Ind}_P^G(\chi^{-1})$ is a Banach space. By Definition 4.13, we have to show that G acts on V by continuous linear automorphisms such that the map $G \times V \to V$ describing the action is continuous.

Fix $g \in G$ and write it as $g = g_0 p$, with $g_0 \in G_0$ and $p \in P$. Set $H = p^{-1} G_0 p$. Then for any $f \in \mathrm{Ind}_P^G(\chi^{-1})$, we have

$$\|L_g(f)\| = \sup_{x \in G_0} |f(p^{-1} g_0^{-1} x)| = \sup_{h \in H} |f(hp^{-1})| = |\chi(p^{-1})| \cdot \|f\|_H.$$

This together with Lemma 7.4 shows that L_g is a continuous linear automorphism on V.

We know that $G_0 \times V \to V$ is continuous. To show that $G \times V \to V$ is continuous, take $g \in G$, $f \in V$, and an open neighborhood U of $L_g f$ in V. By continuity of $G_0 \times V \to V$, there exist a neighborhood U_0 of 1 in G_0 and a neighborhood U_V of $L_g f$ in V such that

$$L_x h \in U \quad \text{for all } x \in U_0, h \in U_V.$$

By continuity of L_g, there exists a neighborhood W of f such that $L_g h \in U_V$ for all $h \in W$. Then for any $y = xg \in U_0 g$ and any $h \in W$ we have $L_y h = L_{xg} h = L_x(L_g h) \in U$, completing the proof. $\qquad\square$

7.1.2 Unitary Principal Series

As defined in Sect. 4.4.1, a Banach space representation (π, V) of G is called unitary if the group action is norm-preserving, that is, $\|\pi(g)v\| = \|v\|$ for all $g \in G$ and $v \in V$. A character $\chi : T \to K^\times$ is unitary if $|\chi(t)| = 1$ for all $t \in T$.

Lemma 7.6 *If* $\chi : T \to K^\times$ *is a continuous unitary character, then* $\mathrm{Ind}_P^G(\chi^{-1})$ *is unitary.*

Proof Since $|\chi(p)| = 1$ for all $p \in P$ and $G = G_0 P$, for any $f \in \mathrm{Ind}_P^G(\chi^{-1})$ we have

$$\|f\| = \sup_{g \in G_0} |f(g)| = \sup_{g \in G} |f(g)|.$$

Then for any $h \in G$ we have $\|L_h f\| = \sup_{g \in G} |f(h^{-1} g)| = \sup_{g \in G} |f(g)| = \|f\|$. $\qquad\square$

7.1.3 Algebraic and Smooth Vectors

Any algebraic character of T is continuous. Also, any smooth character of T is continuous. We want to see how algebraic induction and smooth induction relate to continuous induction.

Algebraic Characters

Let $\chi : T \to K^\times$ be an algebraic character. Suppose χ is dominant. We apply Remark 6.8 to $\mu = \chi^{-1}$. (In the additive notation of Sect. 6.1, we would write $\mu = -\chi$.) Then we have the following algebraic representation

$$\mathrm{ind}_P^G(\chi^{-1}) = \{f : G \to K \text{ algebraic} \mid f(gp) = \chi(p)f(g) \text{ for all } g \in G, p \in P\}.$$

This is a finite-dimensional subspace of $\mathrm{Ind}_P^G(\chi^{-1})$ consisting of all polynomial functions. As a finite-dimensional subspace, $\mathrm{ind}_P^G(\chi^{-1})$ is closed in $\mathrm{Ind}_P^G(\chi^{-1})$. It follows that $\mathrm{Ind}_P^G(\chi^{-1})$ is topologically reducible Banach space representation of G and $\mathrm{ind}_P^G(\chi^{-1})$ is a finite-dimensional G-subrepresentation.

Smooth Characters

Suppose $\chi : T \to K^\times$ is a smooth character. The smooth induced representation $\mathrm{Ind}_P^G(\chi^{-1})^{\mathrm{sm}}$ is usually considered just as an abstract vector space. Since we have a natural embedding

$$\mathrm{Ind}_P^G(\chi^{-1})^{\mathrm{sm}} \subset \mathrm{Ind}_P^G(\chi^{-1}),$$

we can equip $\mathrm{Ind}_P^G(\chi^{-1})^{\mathrm{sm}}$ with the subspace topology.

Recall that $C^\infty(G_0, K)$ is the subspace of $C(G_0, K)$ consisting of smooth (i.e., locally constant) functions. By Exercise 3.20, $C^\infty(G_0, K)$ is dense in $C(G_0, K)$.

Lemma 7.7 *Suppose $\chi : T \to K^\times$ is a smooth character. Then*

(i)

$$\mathrm{Ind}_P^G(\chi^{-1})^{\mathrm{sm}} \cong \bigoplus_{w \in W} C^\infty(U_{w,\frac{1}{2}}^-, K).$$

(ii) $\mathrm{Ind}_P^G(\chi^{-1})^{\mathrm{sm}}$ *is dense in* $\mathrm{Ind}_P^G(\chi^{-1})$.

Proof Any element of $\mathrm{Ind}_P^G(\chi^{-1})$ is a sum over $w \in W$ of elements f_h, where $h \in C(U_{w,\frac{1}{2}}^-, K)$ and f_h is defined as in the proof of Proposition 7.2. If h is smooth, then f_h is also smooth, by smoothness of χ. Since $C^\infty(U^-_{w,\frac{1}{2}}, K)$ is dense in $C(U^-_{w,\frac{1}{2}}, K)$, it follows that $\mathrm{Ind}_P^G(\chi^{-1})^{\mathrm{sm}}$ is dense in $\mathrm{Ind}_P^G(\chi^{-1})$. □

The space $\mathrm{Ind}_P^G(\chi^{-1})^{\mathrm{sm}}$ is G-invariant, but it is not closed in $\mathrm{Ind}_P^G(\chi^{-1})$.

Remark 7.8 There is also the space of locally analytic vectors $\mathrm{Ind}_P^G(\chi^{-1})^{\mathrm{an}}$,

$$\mathrm{Ind}_P^G(\chi^{-1})^{\mathrm{sm}} \subset \mathrm{Ind}_P^G(\chi^{-1})^{\mathrm{an}} \subset \mathrm{Ind}_P^G(\chi^{-1}).$$

Then $\mathrm{Ind}_P^G(\chi^{-1})^{\mathrm{sm}}$ is closed in $\mathrm{Ind}_P^G(\chi^{-1})^{\mathrm{an}}$, under an appropriate topology, and both spaces are dense in $\mathrm{Ind}_P^G(\chi^{-1})$ with respect to the Banach topology [61, Section 3].

7.1.4 Unitary Principal Series of $GL_2(\mathbb{Q}_p)$

Let $G = GL_2(\mathbb{Q}_p)$. Let $P = TU$ be the group of upper triangular matrices, with T the group of diagonal matrices in G and U the unipotent radical of P. Recall the exact sequence of smooth representations

$$0 \to \mathbf{1} \to \mathrm{Ind}_P^G(\mathbf{1})^{\mathrm{sm}} \to \mathrm{St} \to 0.$$

The trivial one-dimensional representation $\mathbf{1}$ is of course closed in $\mathrm{Ind}_P^G(\mathbf{1})$. Denote by $\widehat{\mathrm{St}}$ the quotient $\mathrm{Ind}_P^G(\mathbf{1})/\mathbf{1}$. Then we have the following exact sequence

$$0 \to \mathbf{1} \to \mathrm{Ind}_P^G(\mathbf{1}) \to \widehat{\mathrm{St}} \to 0.$$

We know from Lemma 7.6 that $\mathrm{Ind}_P^G(\mathbf{1})$ is unitary. The representation $\widehat{\mathrm{St}}$ is topologically irreducible and admissible as a representation of G [30, Lemma 5.3.3].

Remark 7.9 Emerton in [29] introduced the notion of a universal unitary Banach space completion. Then $\widehat{\mathrm{St}}$ is the universal unitary completion of St [30, Lemma 5.3.3].

The following proposition describes the reducibility of continuous principal series of $GL_2(\mathbb{Q}_p)$ induced from unitary characters.

Proposition 7.10 *Let* $\chi_1 \otimes \chi_2 : T \to K^\times$ *be a unitary character.*

(i) *If* $\chi_1 \neq \chi_2$, *then* $\mathrm{Ind}_P^G(\chi_1 \otimes \chi_2)$ *is topologically irreducible.*

(ii) *If* $\chi_1 = \chi_2 = \chi$, *then* $\mathrm{Ind}_P^G(\chi \otimes \chi)$ *fits in the following exact sequence*

$$0 \to \chi \circ \det \to \mathrm{Ind}_P^G(\chi \otimes \chi) \to \chi \circ \det \otimes \widehat{\mathrm{St}} \to 0.$$

The above sequence does not split.

Proof This follows from Proposition 5.3.4 in [30]. □

If we compare the above proposition to the smooth case described in Corollary 6.32, we see that both detect reducibility for $\chi_1 = \chi_2$. However, Proposition 7.10 does not cover the case $\chi_1 \chi_2^{-1} = \mid \mid_p^2$ because it treats only unitary characters. Thus, it does not cover the induction from the modulus character δ_P. We will show in Example 8.18 that the representation $\mathrm{Ind}_P^G(\delta_P)$ is topologically irreducible. On the other hand, $\mathrm{Ind}_P^G(\delta_P)^{\mathrm{sm}}$ is reducible, as we know from Corollary 6.32.

7.2   Duals of Principal Series

We will study continuous principal series using the Schneider-Teitelbaum duality described in Chap. 4. Our first step is to describe explicitly duals of principal series (Theorem 7.12).

Let $\chi : P_0 \to o_K^\times$ be a continuous character. By Lemma 4.29 and Corollary 4.30, it extends uniquely to a continuous homomorphism of o_K-modules $\chi : o_K[[P_0]] \to o_K$ and a continuous homomorphism of K-algebras $\chi : K[[P_0]] \to K$. The extension is achieved by $\langle v, \chi \rangle$, where $\langle\ ,\ \rangle : K[[P_0]] \times C(P_0, K) \to K$ is the canonical pairing described in Sect. 3.2.3. Hence, for $v \in K[[P_0]]$ we have

$$\chi(v) = \langle v, \chi \rangle.$$

We denote by $o_K^{(\chi)}$ (respectively, $K^{(\chi)}$) the corresponding one dimensional $o_K[[P_0]]$-module (respectively, $K[[P_0]]$-module).

7.2.1   Module $M_0^{(\chi)}$

With $K^{(\chi)}$ the one dimensional $K[[P_0]]$-module defined above, we define

$$M^{(\chi)} = K[[G_0]] \otimes_{K[[P_0]]} K^{(\chi)} \quad \text{and} \quad M_0^{(\chi)} = o_K[[G_0]] \otimes_{o_K[[P_0]]} o_K^{(\chi)}.$$

Notice that any element of $M^{(\chi)}$ can be written as $\mu \otimes 1$ for some $\mu \in K[[G_0]]$. Hence, $M^{(\chi)}$ is isomorphic to a quotient of $K[[G_0]]$. To describe it explicitly, denote by $N^{(\chi)}$ the left ideal in $K[[G_0]]$ generated by the elements of the form

$$\eta - \chi(\eta), \quad \eta \in K[[P_0]].$$

Then $M^{(\chi)} \cong K[[G_0]]/N^{(\chi)}$. Similarly, $M_0^{(\chi)} \cong o_K[[G_0]]/N_0^{(\chi)}$, where $N_0^{(\chi)}$ is the appropriately defined left ideal in $o_K[[G_0]]$.

Lemma 7.11

(i) If $v \in N^{(\chi)}$ and $f \in \mathrm{Ind}_{P_0}^{G_0}(\chi^{-1})$, then $\langle v, f \rangle = 0$.

(ii) We have a well-defined pairing $\langle \, , \, \rangle : M^{(\chi)} \times \mathrm{Ind}_{P_0}^{G_0}(\chi^{-1}) \to K$ given by

$$\langle \mu \otimes 1, f \rangle = \langle \mu, f \rangle,$$

where the pairing on the right is the canonical pairing $\langle \, , \, \rangle : K[[G_0]] \times C(G_0, K) \to K$.

Proof Assertion (ii) follows from (i). For (i), we have to show that for all $\mu \in K[[G_0]]$, $\eta \in K[[P_0]]$, and $f \in \mathrm{Ind}_{P_0}^{G_0}(\chi^{-1})$,

$$\langle \mu(\eta - \chi(\eta)), f \rangle = 0.$$

To do so, we first observe that for all $g \in G_0$, $p \in P_0$, and $f \in \mathrm{Ind}_{P_0}^{G_0}(\chi^{-1})$,

$$\langle g(p - \chi(p)), f \rangle = f(gp) - \chi(p)f(g) = 0. \tag{7.6}$$

Next, fix $p \in P_0$ and $f \in \mathrm{Ind}_{P_0}^{G_0}(\chi^{-1})$ and define a map $\varphi_{p,f} : K[[G_0]] \to K$ by

$$\varphi_{p,f} : \mu \mapsto \langle \mu(p - \chi(p)), f \rangle.$$

Then $\varphi_{p,f}$ is a continuous K-linear map. Equation (7.6) tells us that the restriction of $\varphi_{p,f}$ to G_0 is zero. Corollary 4.30 then implies $\varphi_{p,f}$ is also zero, so

$$\langle \mu(p - \chi(p)), f \rangle = 0 \tag{7.7}$$

for all $\mu \in K[[G_0]]$, $p \in P_0$, and $f \in \mathrm{Ind}_{P_0}^{G_0}(\chi^{-1})$. Next, fix $\mu \in K[[G_0]]$ and $f \in \mathrm{Ind}_{P_0}^{G_0}(\chi^{-1})$, and define a map $\varphi_{\mu,f} : K[[P_0]] \to K$ by

$$\varphi_{\mu,f}(\eta) = \langle \mu(\eta - \chi(\eta)), f \rangle.$$

This is again a continuous K-linear map and from Corollary 4.30 and Eq. (7.7) we get $\varphi_{\mu,f} = 0$, completing the proof. □

Define

$$m(\chi, n) = \sup\{m \in \mathbb{N} \cup \{0\} \mid \chi(p) \in 1 + \mathfrak{p}_K^m \text{ for all } p \in P_n\},$$

where we follow the convention that $\mathfrak{p}_K^0 = o_K$ and $\mathfrak{p}_K^\infty = 0$. Note that $m(\chi, n) = \infty$ if and only if $\chi|_{P_n} = 1$. In any case, $\lim_{n \to \infty} m(\chi, n) = \infty$ by continuity of χ.

Theorem 7.12 *The continuous dual of* $\mathrm{Ind}_{P_0}^{G_0}(\chi^{-1})$ *is isomorphic to*

$$M^{(\chi)} = K[[G_0]] \otimes_{K[[P_0]]} K^{(\chi)}.$$

Proof Set $V = \mathrm{Ind}_{P_0}^{G_0}(\chi^{-1})$. From the short exact sequence

$$0 \longrightarrow V \longrightarrow C(G_0, K) \longrightarrow C(G_0, K)/V \longrightarrow 0$$

we obtain, taking the continuous dual,

$$0 \longrightarrow H^{(\chi)} \longrightarrow K[[G_0]] \longrightarrow V' \longrightarrow 0,$$

where $H^{(\chi)} = \{\mu \in K[[G_0]] \mid \langle \mu, f \rangle = 0 \,\forall f \in V\}$. Recall that $M^{(\chi)} \cong K[[G_0]]/N^{(\chi)}$. From Lemma 7.11(i), $N^{(\chi)} \subseteq H^{(\chi)}$ and we have a well-defined surjective map

$$\varphi : K[[G_0]]/N^{(\chi)} \longrightarrow V' = K[[G_0]]/H^{(\chi)}$$

$$\varphi(\bar{\mu})(f) = \langle \mu, f \rangle,$$

where $\mu \in K[[G_0]]$ is any representative of $\bar{\mu} \in K[[G_0]]/N^{(\chi)}$ and $f \in V$. To prove that φ is injective, we will show that $N^{(\chi)} = H^{(\chi)}$, and this will be done by showing $N_0^{(\chi)} = H_0^{(\chi)}$, where $H_0^{(\chi)} = H^{(\chi)} \cap o_K[[G_0]]$.

Recall from Proposition 2.50 that $o_K[[G_0]]$ can be expressed as the projective limit

$$o_K[[G_0]] = \varprojlim_{n \in \mathbb{N}} o_K/\mathfrak{p}_K^{m(n)}[G_0/G_n].$$

Let us denote by pr_n the natural projection $\mathrm{pr}_n : o_K[[G_0]] \to o_K/\mathfrak{p}_K^{m(\chi,n)}[G_0/G_n]$. Then

$$H_0^{(\chi)} = \varprojlim_{n \in \mathbb{N}} \mathrm{pr}_n(H_0^{(\chi)}).$$

Define $\Psi_n : N_0^{(\chi)} \to \mathrm{pr}_n(H_0^{(\chi)})$ by $\Psi_n = \mathrm{pr}_n \circ \iota$, where $\iota : N_0^{(\chi)} \hookrightarrow H_0^{(\chi)}$ is the embedding. We claim that Ψ_n is surjective.

To prove the claim, take $\bar{\mu} \in \mathrm{pr}_n(H_0^{(\chi)})$. Then $\bar{\mu} = \mathrm{pr}_n(\mu)$, for some $\mu \in H_0^{(\chi)}$. For $w \in W$, we denote by $H_{w,0}^{(\chi)}$ the set of elements of $H_0^{(\chi)}$ supported on $B\dot{w}B =$

$\dot{w}U^-_{w,\frac{1}{2}}P_0$. Then $H_0^{(\chi)} = \bigoplus_{w \in W} H_{w,0}^{(\chi)}$ and we can write

$$\mu = \sum_{w \in W} \mu_w, \quad \mu_w \in H_{w,0}^{(\chi)}.$$

Then also $\bar{\mu} = \sum_{w \in W} \bar{\mu}_w$, where $\bar{\mu}_w \in \mathrm{pr}_n(H_{w,0}^{(\chi)})$.

Recall the decomposition $G_0 = \coprod_{w \in W} \dot{w}U^-_{w,\frac{1}{2}}P_0$ from Proposition 5.45. Fix $w \in W$. Let $\{u_1, \ldots, u_r\}$ be a set of coset representatives of $U^-_{w,\frac{1}{2}}/U^-_n$ and $\{p_1, \ldots, p_s\}$ a set of coset representatives of P_0/P_n. We can write

$$\bar{\mu}_w = \dot{w} \sum_i \sum_j a_{i,j} u_i p_j G_n,$$

where $a_{i,j} \in o_K/\mathfrak{p}_K^{m(\chi,n)}$, for all i, j. Fix $i \in \{1, \ldots r\}$. Recall the isomorphism $\mathrm{Ind}_{P_0}^{G_0}(\chi_0^{-1})_w \cong C(U^-_{w,\frac{1}{2}}, K)$ form Proposition 7.2. Define $h_i \in C(U^-_{w,\frac{1}{2}}, K)$ to be the characteristic function of $u_i U^-_n$ and let $f_i \in V$ be the function attached to h_i as in the proof of Proposition 7.2. It is defined as

$$f_i(x) = \begin{cases} \chi(p), & x = \dot{w}u_i vp, \ v \in U^-_n, \ p \in P_0 \\ 0, & x \notin \dot{w}u_i U^-_{w,n} P_0. \end{cases}$$

Notice that for $v \neq w$ we have $\langle \mu_v, f_i \rangle = 0$, because $f_i \in \mathrm{Ind}_{P_0}^{G_0}(\chi_0^{-1})_w$ and μ_v is supported on $\dot{v}U^-_{v,\frac{1}{2}}P_0$. Then $\langle \mu, f_i \rangle = \langle \mu_w, f_i \rangle$ and this is equal to zero because $\mu \in H_0^{(\chi)}$.

The function f_i is o_K-valued, and we can define

$$\bar{f}_i = f_i \mod \mathfrak{p}_K^{m(\chi,n)}.$$

We claim that $\bar{f}_i$ is G_n-invariant. If $m(\chi, n) = \infty$, the statement is true because f_i is locally constant. Assume $m(\chi, n) < \infty$. Take $g \in G_n$ and $x \in G_0$. If $x \notin B\dot{w}B$, then $f(g^{-1}x) = 0 = f(x)$. If $x \in B\dot{w}B$, we can write $x = \dot{w}u_j vp$ with $v \in U^-_n$ and $p \in P_0$. Let $h = (\dot{w}u_j v)^{-1}g^{-1}\dot{w}u_j v$. Since G_n is normal in G_0, h is also an element of G_n and we know from Lemma 5.41 that we can write it as $h = v_2 p_2$ where $v_2 \in U^-_n$ and $p_2 \in P_n$. Then

$$L_g f_i(x) = f_i(g^{-1}x) = f_i(\dot{w}u_j vv_2 p_2 p) = \begin{cases} \chi(p_2)\chi(p), & \text{if } j = i, \\ 0, & \text{if } j \neq i. \end{cases}$$

The definition of $m(\chi, n)$ implies that $\chi(p_2) \in 1 + \mathfrak{p}_K^{m(\chi,n)}$. Then

$$L_g f_i(x) \equiv f_i(x) \mod \mathfrak{p}_K^{m(\chi,n)},$$

proving that $\bar{f}_i$ is G_n-invariant. It follows that we have a well-defined $\langle \bar{\mu}_w, \bar{f}_i \rangle \in o_K / \mathfrak{p}_K^{m(\chi,n)}$, and it is equal to zero because $\langle \mu_w, f_i \rangle = 0$. Moreover, since $f_i(u_j) = \delta_{ij}$, we have

$$\langle \bar{\mu}_w, \bar{f}_i \rangle = \sum_j a_{i,j} \bar{\chi}(p_j) = 0,$$

where $\bar{\chi} = \chi \mod \mathfrak{p}_K^{m(\chi,n)}$. This holds for any i. It follows

$$\begin{aligned}
\bar{\mu}_w &= \bar{\mu}_w - \sum_i u_i \left(\sum_j a_{i,j} \bar{\chi}(p_j) \right) G_n \\
&= \sum_i \sum_j a_{i,j} u_i p_j G_n - \sum_i \sum_j a_{i,j} u_i \bar{\chi}(p_j) G_n \\
&= \sum_i \sum_j a_{i,j} u_i (p_j - \bar{\chi}(p_j)) G_n.
\end{aligned}$$

This is clearly an element of $\Psi_n(N_0^{(\chi)})$, thus proving that Ψ_n is surjective.

Hence, we have a family of compatible surjections $\Psi_n : N_0^{(\chi)} \to \mathrm{pr}_n(H_0^{(\chi)})$. By the universal property of projective limits, the corresponding continuous linear map $\Psi : N_0^{(\chi)} \to H_0^{(\chi)}$ is equal to the embedding $\iota : N_0^{(\chi)} \hookrightarrow H_0^{(\chi)}$. Since $N_0^{(\chi)}$ is compact and Ψ_n are surjective, it follows from Corollary 2.20 that $\Psi = \iota$ is surjective. $\qquad\square$

Corollary 7.13 *The continuous principal series* $\mathrm{Ind}_{P_0}^{G_0}(\chi^{-1})$ *is an admissible Banach space representation.*

Proof Since the dual of $\mathrm{Ind}_{P_0}^{G_0}(\chi^{-1})$ is isomorphic to $M^{(\chi)} = K[[G_0]] \otimes_{K[[P_0]]} K^{(\chi)}$, we see that it is finitely generated as a $K[[G_0]]$-module (it is generated by $1 \otimes 1 \in M^{(\chi)}$). By Definition 4.41, it follows that $\mathrm{Ind}_{P_0}^{G_0}(\chi^{-1})$ is admissible. $\qquad\square$

From Proposition 7.3, we have the following decomposition

$$\mathrm{Ind}_P^G(\chi^{-1}) = \bigoplus_{w \in W} \mathrm{Ind}_P^G(\chi^{-1})_w \cong \bigoplus_{w \in W} C(U_{w,\frac{1}{2}}^-, K).$$

Taking dual and applying Theorem 7.12, we get

$$M^{(\chi)} \cong \bigoplus_{w \in W} D^c(U^-_{w,\frac{1}{2}}, K) \cong \bigoplus_{w \in W} K[[U^-_{w,\frac{1}{2}}]]. \tag{7.8}$$

Here, we use the fact that the continuous dual $D^c(U^-_{w,\frac{1}{2}}, K)$ is isomorphic to $K[[U^-_{w,\frac{1}{2}}]]$ (see Theorem 3.44). The isomorphism (7.8) is abstract; we would like to describe the summand $K[[U^-_{w,\frac{1}{2}}]]$ as a subspace of $M^{(\chi)}$. Define

$$M^{(\chi)}_w = \{\mu \otimes 1 \in M^{(\chi)} \mid \langle \mu \otimes 1, f \rangle = 0, \ f \in \operatorname{Ind}^{G_0}_{P_0}(\chi^{-1})_{w'}, \ w' \neq w\} \tag{7.9}$$

and $M^{(\chi)}_{w,0} = M^{(\chi)}_w \cap M^{(\chi)}_0$.

Proposition 7.14

(i) We have a $K[[B]]$-module decomposition

$$M^{(\chi)} = \bigoplus_{w \in W} M^{(\chi)}_w.$$

(ii) For every $w \in W$, define $\varphi_w : K[[U^-_{w,\frac{1}{2}}]] \to M^{(\chi)} = K[[G_0]] \otimes_{K[[P_0]]} K^{(\chi)}$ by $\varphi_w : \mu \mapsto \dot{w}\mu \otimes 1$. Then φ_w induces an isomorphism of K-spaces

$$K[[U^-_{w,\frac{1}{2}}]] \xrightarrow{\sim} M^{(\chi)}_w.$$

Moreover,

$$\varphi = \bigoplus_w \varphi_w : \bigoplus_{w \in W} K[[U^-_{w,\frac{1}{2}}]] \xrightarrow{\sim} M^{(\chi)}$$

is an isomorphism of K-spaces.

(iii) As $K[[V^{\pm}_{w,\frac{1}{2}}]]$-modules, $M^{(\chi)}_w \cong K[[V^{\pm}_{w,\frac{1}{2}}]]$. More specifically, we have an isomorphism

$$\psi_w : K[[V^{\pm}_{w,\frac{1}{2}}]] \xrightarrow{\sim} M^{(\chi)}_w$$

given by $\psi_w : \mu \mapsto \mu\dot{w} \otimes 1$.

Proof

(i) Directly from the definition of $M^{(\chi)}_w$, we obtain $M^{(\chi)} = \bigoplus_{w \in W} M^{(\chi)}_w$, the decomposition of $M^{(\chi)}$ as a K-space. Since each subspace $\operatorname{Ind}^{G_0}_{P_0}(\chi^{-1})_{w'}$ is B-invariant, $M^{(\chi)}_w$ is also B-invariant, and hence a $K[[B]]$-module.

(ii) Any $f \in \mathrm{Ind}_{P_0}^{G_0}(\chi_0^{-1})$ can be written as

$$f = \sum_{w \in W} f_w, \quad f_w \in \mathrm{Ind}_{P_0}^{G_0}(\chi_0^{-1})_w.$$

Each summand f_w is of the form $f_w = f_{h_w}$, where $h_w \in C(U_{w,\frac{1}{2}}^-, K)$ and f_{h_w} is the function attached to h_w as in the proof of Proposition 7.2.

Fix $w \in W$. Then the inclusion $U_{w,\frac{1}{2}}^- \hookrightarrow G_0$ induces an inclusion $o_K[[U_{w,\frac{1}{2}}^-]] \hookrightarrow o_K[[G_0]]$ and hence $K[[U_{w,\frac{1}{2}}^-]] \hookrightarrow K[[G_0]]$. Let $\eta \in K[[U_{w,\frac{1}{2}}^-]]$. We claim that $\langle \dot{w}\eta \otimes 1, f_w \rangle = \langle \eta, h_w \rangle$. Indeed, for general $f \in C(G_0, K)$, the value of $\langle \dot{w}\eta \otimes 1, f \rangle = \langle \dot{w}\eta, f \rangle$ is obtained by applying η to the function $u \mapsto f(\dot{w}u)$, $(u \in U_{w,\frac{1}{2}}^-)$. And, if $f = f_w$ this is precisely h_w.

Moreover, for $v \neq w$, we have $\langle \dot{w}\eta \otimes 1, f_v \rangle = 0$. Combining with (7.9) and (i), we see that for each w, the subspace $\dot{w}K[[U_{w,\frac{1}{2}}^-]] \subset K[[G_0]]$ maps isomorphically onto $M_w^{(\chi)}$. This proves $M_w^{(\chi)} \cong K[[U_{w,\frac{1}{2}}^-]]$. Then also $M_w^{(\chi)} \cong K[[V_{w,\frac{1}{2}}^\pm]]$, because the groups $U_{w,\frac{1}{2}}^-$ and $V_{w,\frac{1}{2}}^\pm$ are conjugate.

(iii) Let us look again at the disjoint unions

$$G_0 = \coprod_{w \in W} \dot{w}U_{w,\frac{1}{2}}^- P_0 = \coprod_{w \in W} V_{w,\frac{1}{2}}^\pm \dot{w} P_0$$

and the direct sum $M^{(\chi)} = \bigoplus_{w \in W} M_w^{(\chi)}$. If $\eta \otimes 1 \in M_w^{(\chi)}$, then there exists a unique $\mu \in K[[U_{w,\frac{1}{2}}^-]]$ such that $\eta = \dot{w}\mu \otimes 1$. Similarly, there exists a unique $\nu \in K[[V_{w,\frac{1}{2}}^\pm]]$ such that $\eta = \nu\dot{w} \otimes 1$. Then $M_w^{(\chi)} \cong K[[V_{w,\frac{1}{2}}^\pm]]$ as $K[[V_{w,\frac{1}{2}}^\pm]]$-modules.

$\square$

Clearly, the isomorphism of K-spaces $\varphi_w : K[[U_{w,\frac{1}{2}}^-]] \to M_w^{(\chi)}$ given by

$$\varphi_w : \mu \mapsto \dot{w}\mu \otimes 1$$

is an isomorphism of $K[[U_{w,\frac{1}{2}}^-]]$-modules only for $w = 1$.

Recall that $M_0^{(\chi)} = o_K[[G_0]] \otimes_{o_K[[P_0]]} o_K^{(\chi)}$. By Theorem 3.32, $\mu \in K[[G_0]]$ lies in $o_K[[G_0]]$ if and only if it maps elements of $C(G_0, o_K) \subset C(G_0, K)$ into o_K. It follows that, for any such μ, the image $\mu \otimes 1 \in M^{(\chi)}$ maps o_K-valued elements of $\mathrm{Ind}_{P_0}^{G_0}(\chi_0^{-1})$ into o_K. Using Proposition 7.14 we show that this characterizes the image of $o_K[[G_0]]$ in $M^{(\chi)}$.

Proposition 7.15

(i) $M_0^{(\chi)}$ is the set of elements of $M^{(\chi)}$ which map o_K-valued elements of $\mathrm{Ind}_{P_0}^{G_0}(\chi^{-1})$ into o_K.

(ii) We have an $o_K[[B]]$-module decomposition

$$M_0^{(\chi)} = \bigoplus_{w \in W} M_{w,0}^{(\chi)}.$$

Proof

(i) We have to prove that any class $[\mu]$ in $M^{(\chi)}$ which maps o_K-valued elements to o_K has a representative in $o_K[[G_0]]$. By Proposition 7.14, we can take a representative of the form $\eta = \sum_{w \in W} \dot{w} \eta_w$ with $\eta_w \in K[[U_{w,\frac{1}{2}}^-]]$. Now, fix w and take $h \in C(U_{w,\frac{1}{2}}^-, K)$. Clearly, the function f_h defined by (7.2) is o_K valued whenever h is. It follows that η_w maps $C(U_{w,\frac{1}{2}}^-, o_K)$ into o_K and hence lies in $o_K[[U_{w,\frac{1}{2}}^-]]$. As this holds for all w and the representatives $\dot{w}$ were taken from G_0, it follows that $\eta \in o_K[[G_0]]$.

(ii) Exercise.

$\square$

Exercise 7.16 Let $\varphi = \bigoplus_w \varphi_w$ be the isomorphism from Proposition 7.14 (iii), where $\varphi_w : \mu \mapsto \dot{w}\mu \otimes 1$. From the embedding $o_K[[G_0]] \hookrightarrow K[[G_0]]$ we obtain

$$f = \bigoplus_w f_w : \bigoplus_{w \in W} o_K[[U_{w,\frac{1}{2}}^-]] \longrightarrow o_K[[G_0]] \otimes_{o_K[[P_0]]} o_K^{(\chi)},$$

where $f_w : \mu \mapsto \dot{w}\mu \otimes 1$. Prove that f is an isomorphism.

The space $M_w^{(\chi)}$ is a $K[[B]]$-module, and so the isomorphism from Proposition 7.14 induces a $K[[B]]$-module structure on $K[[V_{w,\frac{1}{2}}^{\pm}]]$. The action of T_0 can be described explicitly. We start with the standard action of T_0 on $V_{w,\frac{1}{2}}^{\pm}$ (by conjugation), which induces the action of T_0 on $C(V_{w,\frac{1}{2}}^{\pm}, K)$ given by $t \cdot h(u) = h(t^{-1}ut)$. Then, the standard action of T_0 on $K[[V_{w,\frac{1}{2}}^{\pm}]]$ is given by

$$\langle t \cdot \mu, h \rangle = \langle \mu, t^{-1} \cdot h \rangle$$

for $\mu \in K[[V_{w,\frac{1}{2}}^{\pm}]]$, $t \in T_0$, and $h \in C(V_{w,\frac{1}{2}}^{\pm}, K)$. Considering μ and t as elements of $K[[G_0]]$, the above formula implies $t \cdot \mu = t\mu t^{-1}$.

Next, we want to describe the action of T_0 on $K[[V_{w,\frac{1}{2}}^{\pm}]]$ from Proposition 7.14. We use the isomorphism $\psi_w : K[[V_{w,\frac{1}{2}}^{\pm}]] \xrightarrow{\sim} M_w^{(\chi)}$ given by $\psi_w : \mu \mapsto \mu\dot{w} \otimes 1$.

If $t \in T_0$ and $\mu \in K[[V^{\pm}_{w,\frac{1}{2}}]]$, then in $M^{(\chi)}_w \subset M^{(\chi)} = K[[G_0]] \otimes_{K[[P_0]]} K^{(\chi)}$ we have

$$t\mu\dot{w} \otimes 1 = t\mu t^{-1}\dot{w} \otimes (\dot{w}^{-1}t\dot{w}) \cdot 1 = \chi(w^{-1}tw)t\mu t^{-1}\dot{w} \otimes 1 = w\chi(t)t\mu t^{-1}\dot{w} \otimes 1,$$

where $w\chi$ is the character of T_0 defined by $w\chi(t) = \chi(w^{-1}tw)$. Then the action of t on μ is given by $(t, \mu) \mapsto w\chi(t)t\mu t^{-1}$. Let us denote this action by $A_{w\chi}$. It is equal to the standard action twisted by the character $w\chi$. Then $A_{w\chi}(t).\mu = w\chi(t)t \cdot \mu$, and

$$\langle A_{w\chi}(t).\mu, h \rangle = w\chi(t)\langle t \cdot \mu, h \rangle = w\chi(t)\langle \mu, t^{-1} \cdot h \rangle$$

for all $t \in T_0$, $\mu \in K[[V^{\pm}_{w,\frac{1}{2}}]]$, and $h \in C(V^{\pm}_{w,\frac{1}{2}}, K)$. Combined with the action of $K[[V^{\pm}_{w,\frac{1}{2}}]]$ on itself by left translation, this action of T_0 makes $K[[V^{\pm}_{w,\frac{1}{2}}]]$ into a $K[[Q^{\pm}_{w,\frac{1}{2}}]]$-module, where

$$Q^{\pm}_{w,\frac{1}{2}} = T_0 V^{\pm}_{w,\frac{1}{2}} = B \cap wP^{-}_0 w^{-1}.$$

Write $K[[V^{\pm}_{w,\frac{1}{2}}]]^{(w\chi)}$ for this $K[[Q^{\pm}_{w,\frac{1}{2}}]]$-module structure on $K[[V^{\pm}_{w,\frac{1}{2}}]]$. Then we have proved:

Lemma 7.17 As $K[[Q^{\pm}_{w,\frac{1}{2}}]]$-modules, $M^{(\chi)}_w \cong K[[V^{\pm}_{w,\frac{1}{2}}]]^{(w\chi)}$.

7.3   Projective Limit Realization of $M^{(\chi)}_0$

In the rest of the chapter, we follow [4].

In Proposition 7.20, we give a realization of $M^{(\chi)}_0$ as the projective limit over $n \in \mathbb{N}$ of tensor products $o_K[G_0/G_n] \otimes_{o_K[P_0]} o^{(\chi)}_K$. We start by proving two technical lemmas about those tensor products.

Recall that $m(\chi, n) = \sup\{m \in \mathbb{N} \cup \{0\} \mid \chi(p) \in 1 + \mathfrak{p}^m_K$ for all $p \in P_n\}$.

Lemma 7.18 Let $\chi : P_0 \to o^{\times}_K$ be a continuous character and let $n \in \mathbb{N}$.

(i) In $o_K[G_0/G_n] \otimes_{o_K[P_0]} o^{(\chi)}_K$, for any $\xi \in o_K[G_0/G_n]$ and any $b \in \mathfrak{p}^{m(\chi,n)}_K$ we have

$$\xi \otimes b = 0.$$

(ii) The o_K-module $o_K[G_0/G_n] \otimes_{o_K[P_0]} o_K^{(\chi)}$ is isomorphic to

$$\bigoplus_{w \in W} o_K/\mathfrak{p}_K^{m(\chi,n)}[U_{w,\frac{1}{2}}^-/U_n^-]$$

Proof

(i) If $m(\chi, n) = \infty$, then there is nothing to prove.

Assume $m(\chi, n) < \infty$. For any $p \in P_n$ and any $\xi \in o_K[G_0/G_n]$, we have $\xi = \xi p$, and hence

$$\xi \otimes (1 - \chi(p)) = (\xi \otimes 1) - (\xi \otimes \chi(p)) = (\xi \otimes 1) - (\xi p \otimes 1) = 0.$$

Now, take $p_0 \in P_n$ such that $ord_K(\chi(p_0) - 1) = m(\chi, n)$. Then any $b \in \mathfrak{p}_K^{m(\chi,n)}$ can be written as $b = b_0(1 - \chi(p_0))$ for some $b_0 \in o_K$. It follows

$$\xi \otimes b = \xi \otimes b_0(1 - \chi(p_0)) = b_0(\xi \otimes (1 - \chi(p_0))) = 0.$$

(ii) We first recall the disjoint union decomposition $G_0 = \coprod_{w \in W} \dot{w} U_{w,\frac{1}{2}}^- P_0$. Define

$$h_w : o_K[U_{w,\frac{1}{2}}^-/U_n^-] \to o_K[G_0/G_n] \otimes_{o_K[P_0]} o_K^{(\chi)}$$

$$\mu \mapsto \dot{w}\mu \otimes 1.$$

Then $\bigoplus_w h_w : \bigoplus_w o_K[U_{w,\frac{1}{2}}^-/U_n^-] \to o_K[G_0/G_n] \otimes_{o_K[P_0]} o_K^{(\chi)}$ is easily seen to be surjective.

Next, we want to realize $o_K[G_0/G_n] \otimes_{o_K[P_0]} o_K^{(\chi)}$ as the dual of a suitable space of functions. We consider the o_K-module

$$i(\chi, n) := \{f : G_0/G_n \to o_K/\mathfrak{p}_K^{m(\chi,n)} \mid f(gp) = pr_{m(\chi,n)} \chi(p) f(g),$$

$$\text{for } g \in G_0/G_n \text{ and } p \in P_0/P_n\},$$

where $pr_{m(\chi,n)}$ is the canonical projection $o_K \to o_K/\mathfrak{p}_K^{m(\chi,n)}$. The mapping $(g, a) \mapsto \lambda_{g,a}$, where

$$\lambda_{g,a}(f) = af(g), \qquad a \in o_K, \ g \in G_0/G_n, \ f \in i(\chi, n)$$

extends to a surjective $o_K[P_0]$-middle linear map from $o_K[G_0/G_n] \times o_K$ to the o_K-module

$$i(\chi, n)^* := \text{Hom}_{o_K}(i(\chi, n), o_K/\mathfrak{p}_K^{m(\chi,n)}).$$

This middle linear map then induces a linear map $o_K[G_0/G_n] \otimes_{o_K[P_0]}$ $o_K^{(\chi)} \to i(\chi, n)^*$. It is then easy to see that the kernel of the map from $\bigoplus_w w o_K[U_{w,\frac{1}{2}}^-/U_n^-]$ into the $i(\chi, n)^*$ is $\bigoplus_w w \mathfrak{p}_K^{m(\chi,n)}[U_{w,\frac{1}{2}}^-/U_n^-]$.

$\square$

Lemma 7.19 *Let $\mu \in o_K[[G_0]]$ and $v \in o_K[[P_0]]$. Write $\mu = (\mu_n)_{n=1}^{\infty}$ and $v = (v_n)_{n=1}^{\infty}$ as in Sect. 2.3. Then in $o_K[G_0/G_n] \otimes_{o_K[P_0]} o_K^{(\chi)}$ we have*

$$\mu_n v_n \otimes a = \mu_n \otimes \chi(v)a.$$

Proof Let $\{c_n\}_{n=1}^{\infty}$ be a sequence of functions as in Sect. 3.2.3: each $c_n : P_0 \to o_K$ is right P_n-invariant and $\chi = \lim_{n\to\infty} c_n$.

Let us make a reasonable and explicit choice of $\{c_n\}_{n=1}^{\infty}$. For each n, we select $c_n : P_0 \to o_K$ which is constant on cosets of P_n, such that inside each coset there is at least one point p_0 where $c_n(p_0) = \chi(p_0)$.

Now, let $m(\chi, n)$ be the maximal integer such that $\chi(P_n) \subset 1 + \mathfrak{p}_K^{m(\chi,n)}$. If $p_1 P_n = p_2 P_n$, then $\chi(p_1) - \chi(p_2) \in \mathfrak{p}_K^{m(\chi,n)}$. It follows

$$c_n(p) - \chi(p) \in \mathfrak{p}_K^{m(\chi,n)}, \qquad \text{for all } p \in P_0.$$

Consequently, $\langle \xi, c_n - \chi \rangle \in \mathfrak{p}_K^{m(\chi,n)}$ for all $\xi \in o_K[[P_0]]$.

Let $\{p_1, \ldots, p_s\}$ be a set of coset representatives of P_0/P_n consisting of points satisfying $c_n(p_i) = \chi(p_i)$ for all i. (By our construction of c_n, such points exist.) Then we can write $v_n = a_1 p_1 P_n + \cdots + a_s p_s P_n$, where $a_i \in o_K$. Define

$$\eta = a_1 p_1 + \cdots + a_s p_s.$$

This is an element of $o_K[P_0] \subset o_K[[P_0]]$ such that $\eta_n = v_n$. Since

$$\chi(\eta) = a_1\chi(p_1) + \cdots + a_s\chi(p_s) = a_1 c_n(p_1) + \cdots + a_s c_n(p_s) = \langle \eta, c_n \rangle,$$

it follows

$$\chi(\eta) = \langle \eta, c_n \rangle = \langle \eta_n, c_n \rangle = \langle v_n, c_n \rangle = \langle v, c_n \rangle \in \chi(v) + \mathfrak{p}_K^{m(\chi,n)}.$$

Now, in $o_K[G_0/G_n] \otimes_{o_K[P_0]} o_K^{(\chi)}$, we have

$$\mu_n v_n \otimes a = \mu_n \eta_n \otimes a = \mu_n \otimes \chi(\eta)a.$$

To show that the above expression is equal to $\mu_n \otimes \chi(v)a$, we observe that $\chi(\eta) - \chi(v) \in \mathfrak{p}_K^{m(\chi,n)}$, and apply Lemma 7.18(i).

$\square$

Proposition 7.20 *Let* $\chi : P_0 \to o_K^\times$ *be a continuous character trivial on* U_0. *Then*

$$M_0^{(\chi)} \cong \varprojlim_{n \in \mathbb{N}} \left(o_K[G_0/G_n] \otimes_{o_K[P_0]} o_K^{(\chi)} \right).$$

Proof As explained in Sect. 2.3, any $\mu \in o_K[[G_0]]$ can be written as $\mu = (\mu_n)_{n=1}^\infty$, where $\mu_n \in o_K[G_0/G_n]$. For each $n \in \mathbb{N}$, we define a map

$$\psi_n : o_K[[G_0]] \times o_K^{(\chi)} \to o_K[G_0/G_n] \otimes_{o_K[P_0]} o_K^{(\chi)}$$

$$(\mu, a) \mapsto \mu_n \otimes a.$$

It follows from Lemma 7.19 that ψ_n is $o_K[[P_0]]$-middle linear. Hence, it gives rise to a linear map

$$\Psi_n : o_K[[G_0]] \otimes_{o_K[[P_0]]} o_K^{(\chi)} \to o_K[G_0/G_n] \otimes_{o_K[P_0]} o_K^{(\chi)}.$$

Now, $(\Psi_n)_{n \in \mathbb{N}}$ is a family of compatible continuous linear maps which map $o_K[[G_0]] \otimes_{o_K[[P_0]]} o_K^{(\chi)}$ to the inverse system $\left(o_K[G_0/G_n] \otimes_{o_K[P_0]} o_K^{(\chi)} \right)_{n \in \mathbb{N}}$. By the universal property of projective limits, there exists a continuous linear map

$$\Psi : M_0^{(\chi)} = o_K[[G_0]] \otimes_{o_K[[P_0]]} o_K^{(\chi)} \to \varprojlim_{n \in \mathbb{N}} \left(o_K[G_0/G_n] \otimes_{o_K[P_0]} o_K^{(\chi)} \right).$$

This map is surjective because $M_0^{(\chi)} = o_K[[G_0]] \otimes_{o_K[[P_0]]} o_K^{(\chi)}$ is compact and Ψ_n are surjective (Corollary 2.20).

For injectivity, we first recall from Proposition 7.14 and Exercise 7.16 the isomorphism

$$f = \bigoplus_w f_w : \bigoplus_{w \in W} o_K[[U_{w,\frac{1}{2}}^-]] \xrightarrow{\sim} o_K[[G_0]] \otimes_{o_K[[P_0]]} o_K^{(\chi)},$$

where $f_w : \mu \mapsto \dot{w}\mu \otimes 1$ for $\mu \in o_K[[U_{w,\frac{1}{2}}^-]]$.

For every $w \in W$, we have the following commutative diagram

$$
\begin{array}{ccc}
o_K[[U_{w,\frac{1}{2}}^-]] & \xrightarrow{h_w} & \varprojlim_{n \in \mathbb{N}} \left(o_K/\mathfrak{p}_K^{m(\chi,n)}[U_{w,\frac{1}{2}}^-/U_n^-] \right) \\
\downarrow{\scriptstyle f_w} & & \downarrow{\scriptstyle g_w} \\
o_K[[G_0]] \otimes_{o_K[[P_0]]} o_K^{(\chi)} & \xrightarrow[\Psi]{} & \varprojlim_{n \in \mathbb{N}} \left(o_K[G_0/G_n] \otimes_{o_K[P_0]} o_K^{(\chi)} \right).
\end{array}
$$

The map h_w is built from the natural projections

$$o_K[[U^-_{w,\frac{1}{2}}]] \to o_K/\mathfrak{p}^{m(\chi,n)}_K[U^-_{w,\frac{1}{2}}/U^-_n],$$

using the universal property of projective limits. The map g_w is defined as follows. We know from the proof of Lemma 7.18(ii) that the maps $g_{n,w}$: $o_K/\mathfrak{p}^{m(\chi,n)}_K[U^-_{w,\frac{1}{2}}/U^-_n] \to o_K[G_0/G_n]\otimes_{o_K[P_0]}o^{(\chi)}_K$, given by $g_{n,w} : \mu \mapsto \dot{w}\mu \otimes 1$, are injective, and that $g_n = \bigoplus_w g_{n,w}$ is an isomorphism of o_K-modules. Define $g_w = \varprojlim_n g_{n,w}$. Then g_w is injective. Thus, we reduce our proof to proving the injectivity of h_w, for all $w \in W$.

Suppose η is a nonzero element of $o_K[[U^-_{w,\frac{1}{2}}]]$ and write $\eta = (\eta_n)^{\infty}_{n=1}$ where $\eta_n \in o_K[U^-_{w,\frac{1}{2}}/U^-_n]$. Then for each n we have

$$\eta_n = \sum_{\overline{u} \in U^-_{w,\frac{1}{2}}/U^-_n} c_{\overline{u}}\overline{u}$$

and for some $n_0, \overline{u}_0\ c_{\overline{u}_0} \neq 0$. Then for all $n \geq n_0$ there exists $\overline{u} \in U^-_{w,\frac{1}{2}}/U^-_n$ such that $|c_{\overline{u}}| \geq |c_{\overline{u}_0}|$. Then for all n sufficiently large we will have $c_{\overline{u}_0} \notin \mathfrak{p}^{m(n,\chi)}_K$, and hence the image of η_n in $o_K/\mathfrak{p}^{m(\chi,n)}_K[U^-_{w,\frac{1}{2}}/U^-_n]$ is nonzero. $\square$

7.4 Direct Sum Decomposition of $M^{(\chi)}$

Recall the space $\mathrm{Ind}^{G_0}_{P_0}(\chi^{-1})_w = \{f \in \mathrm{Ind}^{G_0}_{P_0}(\chi^{-1}) \mid \mathrm{supp}(f) \subset B\dot{w}B\}$, and its dual $M^{(\chi)}_w$. The purpose of this section is to give a realization of $M^{(\chi)}_w$ as a tensor product, analogous to the realization of $M^{(\chi)}$ itself as $K[[G_0]] \otimes_{K[[P_0]]} K^{(\chi_0)}$. It is motivated by the case of $G_0 = GL_2(\mathbb{Z}_p)$ described by Schneider and Teitelbaum in [64].

7.4.1 The Case $G_0 = GL_2(\mathbb{Z}_p)$

Let $\mathbf{G} = GL_2$. As in Chap. 5, we denote by $\mathbf{P}$ the Borel subgroup of upper triangular matrices, with Levi decomposition $\mathbf{P} = \mathbf{T}\mathbf{U}$. The opposite parabolic subgroup $\mathbf{P}^- = \mathbf{T}\mathbf{U}^-$ consists of lower triangular matrices. The Weyl group $W = W(\mathbf{G}, \mathbf{T}) = \{1, w\}$, and we select the representative $\dot{w} = \begin{pmatrix} 0 & 1 \\ 1 & 0 \end{pmatrix}$. Let $G_0 = GL_2(\mathbb{Z}_p)$. Let B be the Iwahori subgroup of all matrices which are upper triangular modulo p. Then B

decomposes as $B = U_1^- T_0 U_0$. Let $\chi : T_0 \to o_K^\times$ be a continuous character. Define

$$N_\chi = K[[B]] \otimes_{K[[P_0]]} K^{(\chi)} \quad \text{and} \quad N_{w\chi}^- = K[[B]] \otimes_{K[[T_0 U_1^-]]} K^{(w\chi)}.$$

Then from [64, Section 4] we have the following isomorphism of $K[[B]]$-modules

$$M^{(\chi)} \cong N_\chi \oplus N_{w\chi}^-$$

The statement follows from the disjoint union decomposition $G_0 = B \bigsqcup B\dot{w} P_0$ and the proof is left as an exercise. We will prove a general statement for an arbitrary **G** (see Corollary 7.24).

Exercise 7.21 Let $G_0 = GL_2(\mathbb{Z}_p)$. Prove $M^{(\chi)} \cong N_\chi \oplus N_{w\chi}^-$.

7.4.2 General Case

We will prove that $\mathrm{Ind}_{P_0}^{G_0}(\chi^{-1})_w$ is isomorphic as a B-module to a representation induced from $B \cap \dot{w} P_0 \dot{w}^{-1}$, and obtain the corresponding tensor product expression for $M_w^{(\chi)}$. To prepare for the proof, we introduce the following technical result.

Lemma 7.22 Let F be any field. Let $\Phi^+ = S_1 \bigsqcup S_2$ be any partition of the positive roots into two disjoint sets. Take any numbering of S_1 as $\{\beta_1, \ldots, \beta_n\}$ and any numbering of S_2 as $\{\gamma_1, \ldots, \gamma_m\}$. Then

$$\Big((b_1, \ldots, b_n), \ t, \ (c_1, \ldots, c_m)\Big) \mapsto x_{\beta_1}(b_1) \ldots x_{\beta_n}(b_n) \cdot t \cdot x_{\gamma_1}(c_1) \ldots x_{\gamma_m}(c_m)$$

is a bijection $F^n \times \mathbf{T}(F) \times F^m \to \mathbf{P}(F)$.

Proof By §14.4 of [8], multiplying root subgroups gives an isomorphism of varieties $\prod_\alpha \mathbf{U}_\alpha \to \mathbf{U}$, for any ordering of the roots. Since **G** is $\mathbb{Z}$-split,

$$((b_1, \ldots, b_n), (c_1, \ldots, c_n)) \to x_{\beta_1}(b_1) \ldots x_{\beta_n}(b_n) x_{\gamma_1}(c_1) \ldots x_{\gamma_m}(c_m)$$

is a bijection $F^n \times F^m \to U$. On the other hand, $P = TU$, and we can conjugate $t \in T$ to the middle. □

Recall the decomposition $G_0 = \bigsqcup_{w \in W} B\dot{w}B = \bigsqcup_{w \in W} V_{w,\frac{1}{2}}^{\pm} \dot{w} P_0$, where

$$V_{w,\frac{1}{2}}^{\pm} = B \cap \dot{w} U_0^- \dot{w}^{-1} = (U_0 \cap \dot{w} U^- \dot{w}^{-1})(U_1^- \cap \dot{w} U^- \dot{w}^{-1}).$$

Its "multiplicative complement" in B is

$$P_{\frac{1}{2}}^{w,\pm} = B \cap \dot{w} P_0 \dot{w}^{-1} = T_0 (U_0 \cap \dot{w} U_0 \dot{w}^{-1})(U_1^- \cap \dot{w} U_0 \dot{w}^{-1}),$$

as we will show in Lemma 7.23 below. The unipotent radical of $P_{\frac{1}{2}}^{w,\pm}$ is

$$V_{\frac{1}{2}}^{w,\pm} = B \cap \dot{w} U_0 \dot{w}^{-1} = (U_0 \cap \dot{w} U_0 \dot{w}^{-1})(U_1^- \cap \dot{w} U_0 \dot{w}^{-1}).$$

Lemma 7.23

(i) *Multiplication induces a homeomorphism* $V_{w,\frac{1}{2}}^{\pm} \times P_{\frac{1}{2}}^{w,\pm} \to B.$

(ii) *As representations of B,*

$$\mathrm{Ind}_{P_0}^{G_0}(\chi^{-1})_w \cong \mathrm{Ind}_{P_{\frac{1}{2}}^{w,\pm}}^{B} w\chi^{-1}.$$

(iii) *As $K[[B]]$-modules,*

$$M_w^{(\chi)} \cong K[[B]] \otimes_{K[[P_{\frac{1}{2}}^{w,\pm}]]} K^{(w\chi)}.$$

Proof

(i) We know that the multiplication induces an isomorphism of varieties $\mathbf{U}^- \times \mathbf{P} \xrightarrow{\sim} \mathbf{U}^- \mathbf{P}$. Conjugating with w, we get

$$w\mathbf{U}^- w^{-1} \times w\mathbf{P}w^{-1} \xrightarrow{\sim} (w\mathbf{U}^- w^{-1})(w\mathbf{P}w^{-1}).$$

Let us look at the o_L-points contained in B. Recall that we write $\bar{P}$ for $P_0/P_1 = \mathbf{P}(o_L/\mathfrak{p}_L) = B/G_1$. Similarly, we define $\bar{T}, \bar{U}$ and $\bar{U}_\alpha$ for each root α. Given $b \in B$, let $\bar{b}$ be the image in $\bar{P}$. We factor it as

$$\bar{b} = \bar{u}_1 \bar{t} \bar{u}_2, \quad \bar{u}_1 \in \prod_{\alpha > 0,\ w^{-1}\alpha < 0} U_\alpha(o_L/\mathfrak{p}_L), \quad \bar{u}_2 \in \prod_{\alpha > 0,\ w^{-1}\alpha > 0} U_\alpha(o_L/\mathfrak{p}_L).$$

Choosing representatives in $U_\alpha(o_L)$ and $T(o_L)$, we obtain u_1, u_2 and t such that $g_1 := u_1^{-1} b u_2^{-1} t^{-1} \in G_1 \subset B$. Now using the Iwahori factorization of B we can write $\dot{w}^{-1} g_1 \dot{w}$ as $v_1 v_2$ with $v_1 \in U_1^-$ and $v_2 \in P_1 = T_1 U_1$. Put $v_i' = \dot{w} v_i \dot{w}^{-1}$. Then $g_1 = v_1' v_2'$ with $v_1' \in \dot{w} U_1^- \dot{w}^{-1}$ and $v_2' \in T_1 \dot{w} U_1 \dot{w}^{-1}$. Now $b = u_1 v_1' v_2' t u_2$, and $u_1 v_1' \in V_{w,\frac{1}{2}}^{\pm}$ and $v_2' t u_2 \in P_{\frac{1}{2}}^{w,\pm}$.

(ii) For $f \in (\mathrm{Ind}_{P_0}^{G_0} \chi^{-1})_w$, define $\varphi(f) : B \to K$ by $\varphi(f)(b) = f(b\dot{w})$. Then for $b \in B$ and $p \in P_{\frac{1}{2}}^{w,\pm}$, we have

$$\varphi(f)(bp) = f(bp\dot{w}) = f(b\dot{w}\dot{w}^{-1}p\dot{w}) = \chi(\dot{w}^{-1}p\dot{w})f(b\dot{w}) = w\chi(p)\varphi(f),$$

because $\dot{w}^{-1}p\dot{w} \in P_0$ by definition of $P_{\frac{1}{2}}^{w,\pm}$. This proves that $\varphi(f) \in \mathrm{Ind}_{P_{\frac{1}{2}}^{w,\pm}}^{B}(w\chi^{-1})$. The resulting map

$$\varphi : (\mathrm{Ind}_{P_0}^{G_0} \chi^{-1})_w \to \mathrm{Ind}_{P_{\frac{1}{2}}^{w,\pm}}^{B}(w\chi^{-1})$$

is clearly linear and injective. To show that it is surjective, we use the homeomorphism $V_{w,\frac{1}{2}}^{\pm} \times P_{\frac{1}{2}}^{w,\pm} \to B$. Then $\mathrm{Ind}_{P_{\frac{1}{2}}^{w,\pm}}^{B}(w\chi^{-1}) \cong C(V_{w,\frac{1}{2}}^{\pm}, K)$, with $f \in \mathrm{Ind}_{P_{\frac{1}{2}}^{w,\pm}}^{B}(w\chi^{-1})$ given by

$$f(vp) = h(v)w\chi(p), \qquad (v \in V_{w,\frac{1}{2}}^{\pm}, p \in P_{\frac{1}{2}}^{w,\pm}),$$

for some $h \in C(V_{w,\frac{1}{2}}^{\pm}, K)$. From this, it easily follows that φ is surjective. Finally, we see directly from the definition that φ is B-equivariant under left translations, thus proving that φ is an isomorphism of B-representations.

(iii) It follows readily from the definitions that $\langle \mu\pi, f \rangle = w\chi(\pi)\langle \mu, f \rangle$ for all $\mu \in K[[B]]$, $\pi \in K[[P_{\frac{1}{2}}^{w,\pm}]]$, and $f \in \mathrm{Ind}_{P_{\frac{1}{2}}^{w,\pm}}^{B}(w\chi^{-1})$. It follows that the map

$(\mu, a) \to a\mu\Big|_{\mathrm{Ind}_{P_{\frac{1}{2}}^{w,\pm}}^{B}(w\chi^{-1})}$ is a middle-linear map from $K[[B]] \times K^{(w\chi)}$ to the

dual of $\mathrm{Ind}_{P_{\frac{1}{2}}^{w,\pm}}^{B}(w\chi^{-1})$, and hence gives rise to a map from $K[[B]] \otimes_{K[[P_{\frac{1}{2}}^{w,\pm}]]} K^{(w\chi)}$ to this dual.

Since $\mathrm{Ind}_{P_{\frac{1}{2}}^{w,\pm}}^{B}(w\chi^{-1}) \cong C(V_{w,\frac{1}{2}}^{\pm}, K)$, it follows that $K[[V_{w,\frac{1}{2}}^{\pm}]]$ maps isomorphically onto the dual, and from this it follows that the map from $K[[B]] \otimes_{K[[P_{\frac{1}{2}}^{w,\pm}]]} K^{(w\chi)}$ is surjective.

The same reasoning used in Proposition 7.14 to show that $\dot{w}K[[U_{w,\frac{1}{2}}^{-}]]$ surjects onto $M_w^{(\chi)}$ may be used here to prove that $K[[V_{w,\frac{1}{2}}^{\pm}]]$ surjects onto $K[[B]] \otimes_{K[[P_{\frac{1}{2}}^{w,\pm}]]} K^{(w\chi)}$. Since the map from $K[[V_{w,\frac{1}{2}}^{\pm}]]$ to the dual of

$\mathrm{Ind}_{P_{\frac{1}{2}}^{w,\pm}}^{B}(w\chi^{-1})$ is an isomorphism, the map from $K[[B]] \otimes_{K[[P_{\frac{1}{2}}^{w,\pm}]]} K^{(w\chi)}$ onto the dual of $\mathrm{Ind}_{P_{\frac{1}{2}}^{w,\pm}}^{B}(w\chi^{-1})$ must be injective, which completes the proof.

$\square$

Corollary 7.24 *As $K[[B]]$-modules,*

$$M^{(\chi)} \cong \bigoplus_{w \in W} K[[B]] \otimes_{K[[P_{\frac{1}{2}}^{w,\pm}]]} K^{(w\chi)}.$$

Proof From Proposition 7.14, we have a $K[[B]]$-module decomposition $M^{(\chi)} = \bigoplus_{w \in W} M_{w}^{(\chi)}$. Lemma 7.23(iii) then gives the statement. $\square$

Chapter 8
Intertwining Operators

Throughout Part II, $\mathbb{Q}_p \subseteq L \subseteq K$ is a sequence of finite extensions, $\mathbf{G}$ is a split connected reductive $\mathbb{Z}$-group and $\mathbf{P} = \mathbf{TU}$ is a Borel subgroup of $\mathbf{G}$. Furthermore, $G = \mathbf{G}(L)$ and $G_0 = \mathbf{G}(o_L)$. Additional notation is listed on page 89.

In Sects. 8.1 and 8.2, we present the main results and proofs from [4]. The purpose is to describe the space of continuous G_0-intertwining operators

$$\mathrm{Hom}^c_{G_0}(\mathrm{Ind}^{G_0}_{P_0}(\chi_1), \mathrm{Ind}^{G_0}_{P_0}(\chi_2))$$

for two continuous characters $\chi_1, \chi_2 : T_0 \to o_L^\times$. As before, we apply the Schneider-Teitelbaum duality and work with the corresponding $K[[G_0]]$-modules $M^{(\chi_1)}$ and $M^{(\chi_2)}$. We show that

$$\mathrm{Hom}_{K[[G_0]]}(M^{(\chi_1)}, M^{(\chi_2)}) = \begin{cases} 0 & \text{if } \chi_1 \neq \chi_2, \\ K \cdot \mathrm{id} & \text{if } \chi_1 = \chi_2, \end{cases}$$

(see Corollary 8.13). By duality, this gives us

$$\mathrm{Hom}^c_{G_0}(\mathrm{Ind}^{G_0}_{P_0}(\chi_1), \mathrm{Ind}^{G_0}_{P_0}(\chi_2)) = \begin{cases} 0 & \text{if } \chi_1 \neq \chi_2, \\ K \cdot \mathrm{id} & \text{if } \chi_1 = \chi_2, \end{cases}$$

(Corollary 8.14). So, in the world of continuous principal series on K-Banach spaces, the intertwining operators are essentially non-existent: there are no intertwiners between different principal series, and only possible self-intertwining operators are scalar multiples of the identity map. This stands in striking contrast to the world of smooth principal series, where intertwiners are abundant.

As a consequence, the question of reducibility of continuous principal series cannot be solved using intertwining operators. In Sect. 8.4, we present some methods for addressing this question. In Sect. 8.4.1, we state Schneider's conjecture

© The Author(s), under exclusive license to Springer Nature Switzerland AG 2022
D. Ban, *p-adic Banach Space Representations*, Lecture Notes
in Mathematics 2325, https://doi.org/10.1007/978-3-031-22684-7_8

and describe some irreducibility results which come from locally analytic vectors. In Sect. 8.4.2, we present the criterion for irreducibility from [3].

8.1 Invariant Distributions

In this section, we will prove that $\mathrm{Hom}_{K[[P_0]]}(K^{(\chi_1)}, M_w^{(\chi_2)}) = 0$ for all w other than the identity. In fact, what we prove in Proposition 8.4 is a more general statement which allows us, in Corollary 8.6 to show that $\mathrm{Hom}_{K[[B]]}(M_w^{(\chi_1)}, M_v^{(\chi_2)}) \neq 0$ implies $w = v$.

8.1.1 Invariant Distributions on Vector Groups

Recall that the space of continuous distributions $D^c(G_0, K)$ is isomorphic to $K[[G_0]]$. In this section, we will study distributions on unipotent subgroups. We will prove that the only invariant distribution on a group isomorphic to several copies of $\mathbb{Z}_p$ is the trivial one. If $\mathbf{N}$ is an abelian unipotent group, this applies to the groups $N_n, n \in \mathbb{N}$. These are the "vector groups" of the title.

Lemma 8.1 *Suppose* $V \cong o_L^r$ *for some positive integer* r *and that* $\mu \in K[[V]]$ *satisfies*

$$v \cdot \mu = \mu, \qquad \forall v \in V.$$

Then $\mu = 0$. *That is, the space* $K[[V]]^V$ *of* V-*invariant distributions on* V *is* 0.

Proof Let $V_0 = V$ and for $n = 1, 2, 3 \ldots$ let V_n be the image of the r-copies of $\mathfrak{p}_L^n$ under an isomorphism $o_L^r \to V$. So, $[V_m : V_n] = q_L^{r(n-m)}$ for any nonnegative integers m, n with $m < n$. Now, there is a constant c such that $|\mu(f)| \leq c$ for all $f \in C(V, o_K)$. This follows from the fact that $\mu = a\mu_0$ for some $a \in K$ and $\mu_0 \in o_K[[G_0]]$, and $|\mu_0(f)| \leq 1$ for all $f \in C(V, o_K)$. Then

$$|\mu(\mathbf{1}_{V_m})| = |q_L^{r(n-m)}\mu(\mathbf{1}_{V_n})| \leq c|q_L^{r(n-m)}|$$

for all n, m. Since $c|q_L^{r(n-m)}| \to 0$ as $n \to \infty$ for each fixed m, we deduce that $\mu(\mathbf{1}_{V_m}) = 0$ for all m. But the space spanned by translates of these functions is $C^\infty(V, K)$ and, as observed in Sect. 3.2.3, it is dense in $C(V, K)$.   □

Corollary 8.2 *Take* V *as in Lemma 8.1. Regard* K *as a* $K[[V]]$ *module with trivial action. Then*

$$\mathrm{Hom}_{K[[V]]}(K, K[[V]]) = 0.$$

Proof The image of any element of $\mathrm{Hom}_{K[[V]]}(K, K[[V]])$ is an element of $K[[V]]^V$. $\qquad\square$

8.1.2 *"Partially Invariant" Distributions on Unipotent Groups*

Lemma 8.3 *Let $V_0 \subset G_0$ be a subgroup. Let V_1 be a closed subgroup of V_0 which is isomorphic to o_L^r for some r, and V_2 a closed subset of V_0 such that multiplication is a homeomorphism $V_1 \times V_2 \to V_0$. Then $K[[V_0]]^{V_1} = 0$ and hence*

$$\mathrm{Hom}_{K[[V_1]]}(K, K[[V_0]]) = 0.$$

Proof Since multiplication is a homeomorphism $V_1 \times V_2 \to V_0$, we have an injective map $C(V_1, K) \times C(V_2, K) \hookrightarrow C(V_0, K)$. For each fixed nonzero $h \in C(V_2, K)$ we get an injective map $i_h : C(V_1, K) \hookrightarrow C(V_0, K)$. Explicitly,

$$i_h(f)(uv) = f(u)h(v), \qquad (u \in V_1, \ v \in V_2).$$

Assume that $\mu \in K[[V_0]]$ is a V_1-invariant element. Then $\mu \circ i_h$ is an invariant element of $K[[V_1]]$. Thus $\mu \circ i_h = 0$, by Lemma 8.1. But this means that μ vanishes on $i_h.f$ for all f and all h. That is μ vanishes on the image of $C(V_1, K) \times C(V_2, K)$ in $C(V_0, K)$. But the span of this image is dense, so μ must vanish identically. $\qquad\square$

Proposition 8.4 *Suppose $\chi, \xi : T_0 \to o_K^\times$ are continuous characters (we allow $\chi = \xi$). If $w, v \in W$, $w \neq v$, then*

$$\mathrm{Hom}_{K[[P_{\frac{1}{2}}^{w,\pm}]]}(K^{(w\chi)}, M_v^{(\xi)}) = 0.$$

Proof With $V_{\frac{1}{2}}^{w,\pm}$ as in Sect. 7.4.2, we have $P_{\frac{1}{2}}^{w,\pm} = T_0 V_{\frac{1}{2}}^{w,\pm}$. We select a root γ such that $w^{-1}\gamma > 0$ and $v^{-1}\gamma < 0$, and define

$$\epsilon = \begin{cases} 0, & \text{if } \gamma > 0, \\ 1, & \text{if } \gamma < 0. \end{cases}$$

Then $U_{\gamma,\epsilon} \subset V_{\frac{1}{2}}^{w,\pm} \cap V_{v,\frac{1}{2}}^{\pm}$. Clearly

$$\mathrm{Hom}_{K[[P_{\frac{1}{2}}^{w,\pm}]]}(K^{(w\chi)}, M_v^{(\xi)}) \subset \mathrm{Hom}_{K[[U_{\gamma,\epsilon}]]}(K, M_v^{(\xi)}).$$

But $M_v^{(\xi)} \cong K[[V_{v,\frac{1}{2}}^{\pm}]]$ as a $K[[V_{v,\frac{1}{2}}^{\pm}]]$ and hence as a $K[[U_{\gamma,\epsilon}]]$-module, and it follows from Lemma 8.3 that $\mathrm{Hom}_{K[[U_{\gamma,\epsilon}]]}(K, K[[V_{v,\frac{1}{2}}^{\pm}]]) = 0$. $\qquad \square$

In the proof of Corollary 8.6, we use the following result from abstract algebra:

Theorem 8.5 (Adjoint Isomorphism) *Suppose R and S are rings with unity.*

(i) If A is a left R-module, B is an (S, R)-bimodule, and C is a left S-module, then

$$\mathrm{Hom}_S(B \otimes_R A, C) \cong \mathrm{Hom}_R(A, \mathrm{Hom}_S(B, C)).$$

(ii) If A is a right R-module, B is an (R, S)-bimodule, and C is a right S-module, then

$$\mathrm{Hom}_S(A \otimes_R B, C) \cong \mathrm{Hom}_R(A, \mathrm{Hom}_S(B, C)).$$

Proof Rotman [56], Theorem 2.11. $\qquad \square$

Corollary 8.6 *Suppose $\chi, \xi : T_0 \to o_K^\times$ are continuous characters (we allow $\chi = \xi$). If $w, v \in W$, $w \neq v$, then*

$$\mathrm{Hom}_{K[[B]]}(M_w^{(\chi)}, M_v^{(\xi)}) = 0.$$

Proof Since $M_w^{(\chi)} \cong K[[B]] \otimes_{K[[P_{\frac{1}{2}}^{w,\pm}]]} K^{(w\chi)}$, we have

$$\mathrm{Hom}_{K[[B]]}(M_w^{(\chi)}, M_v^{(\xi)}) \cong \mathrm{Hom}_{K[[B]]}(K[[B]] \otimes_{K[[P_{\frac{1}{2}}^{w,\pm}]]} K^{(w\chi)}, M_v^{(\xi)}).$$

We can regard $K[[B]]$ as a bimodule with $K[[B]]$ acting on the left and $K[[P_{\frac{1}{2}}^{w,\pm}]]$ acting on the right, and apply adjoint associativity (Theorem 8.5(i)). It follows that

$$\mathrm{Hom}_{K[[B]]}(K[[B]] \otimes_{K[[P_{\frac{1}{2}}^{w,\pm}]]} K^{(w\chi)}, M_v^{(\xi)})$$

$$\cong \mathrm{Hom}_{K[[P_{\frac{1}{2}}^{w,\pm}]]}(K^{(w\chi)}, \mathrm{Hom}_{K[[B]]}(K[[B]], M_v^{(\xi)})).$$

And $\mathrm{Hom}_{K[[B]]}(K[[B]], M_v^{(\xi)})$ has the structure of a left $K[[P_{\frac{1}{2}}^{w,\pm}]]$-module isomorphic to $M_v^{(\xi)}$ by Theorems 1.15 and 1.16 of [56], so

$$\mathrm{Hom}_{K[[P_{\frac{1}{2}}^{w,\pm}]]}(K^{(w\chi)}, \mathrm{Hom}_{K[[B]]}(K[[B]], M_v^{(\xi)})) \cong \mathrm{Hom}_{K[[P_{\frac{1}{2}}^{w,\pm}]]}(K^{(w\chi)}, M_v^{(\xi)}),$$

which is zero by Proposition 8.4. $\qquad \square$

8.1.3 T_0-Equivariant Distributions on Unipotent Groups

Let $\mathbf{V}$ be a $\mathbf{T}$-stable unipotent $\mathbb{Z}$-subgroup of $\mathbf{G}$. The natural action of T on V by conjugation induces actions of T_0 on V_n/V_m for all positive integers m, n with $m > n$.

To describe T_0-equivariant distributions on V, we will consider their canonical pairing with characteristic functions of the form $1_{u_0 V_n}$. Fix $n > 0$ and $u_0 \in V_0$ such that $u_0 \notin V_n$. We will further decompose $1_{u_0 V_n}$ as the sum of characteristic functions of the form $1_{u V_m}$, for $m > n$. To simplify notation, we define, for $m > n$,

$$Q_m = Q_m^{u_0, n} = \{u_0 v V_m \mid v \in V_n\}.$$

This is a subset of the quotient group V_0/V_m.

Lemma 8.7 *The set $u_0 V_n = \{u_0 v \mid v \in V_n\}$ is preserved under the action of T_n. Consequently, the set Q_m is also preserved under the action of T_n.*

Proof From the hypothesis that $\mathbf{V}$ is $\mathbf{T}$-stable, we deduce that there is a set S of roots such that taking any fixed order on the elements of S and multiplying in that order gives an isomorphism of $\mathbb{Z}$-schemes $\prod_{\alpha \in S} \mathbf{U}_\alpha \to \mathbf{V}$. (Cf. [40, §1.7, p. 159]) In particular, we get a homeomorphism $\prod_{\alpha \in S} U_\alpha(o_L) \to V(o_L)$, which induces a bijection $\prod_{\alpha \in S} U_\alpha(o_L/\mathfrak{p}_L^n) \to V(o_L/\mathfrak{p}_L^n)$, for each n, from which we deduce that the preimage of V_n in $\prod_{\alpha \in S} U_\alpha$ is $\prod_{\alpha \in S} U_{\alpha, n}$.

Hence, we can write u_0 as

$$u_0 = \prod_{\alpha \in S} x_\alpha(r_\alpha),$$

where $r_\alpha \in o_L$. Let $t \in T_n$. For each root $\alpha \in S$

$$t x_\alpha(r_\alpha) t^{-1} = x_\alpha(r_\alpha t^\alpha) = x_\alpha(r_\alpha) x_\alpha(r_\alpha(t^\alpha - 1)).$$

But $t^\alpha \equiv 1 \pmod{\mathfrak{p}_L^n}$, so $x_\alpha(r_\alpha(t^\alpha - 1)) \in V_n$. $\square$

Given $\bar{u} \in Q_m$, we denote by $\mathrm{Orb}_{T_n}(\bar{u})$ its orbit under the action of T_n. Then $\mathrm{Orb}_{T_n}(\bar{u}) \subset Q_m \subset V_0/V_m$ is a finite set and we denote its cardinality by $|\mathrm{Orb}_{T_n}(\bar{u})|$.

Lemma 8.8 *Let $\mathbf{V}$ be a $\mathbf{T}$-stable unipotent $\mathbb{Z}$-subgroup of $\mathbf{G}$. Take $u_0 \in V_0$ and $n > 0$ such that $u_0 \notin V_n$. Then*

$$\lim_{m \to \infty} \min_{\bar{u} \in Q_m} ord_p(|\mathrm{Orb}_{T_n}(\bar{u})|) = \infty,$$

where ord_p is the p-adic valuation on $\mathbb{Z}$.

Proof Let $\bar{u} = u_0 v V_m \in Q_m$ and write $u_0 v = \prod_{\alpha \in S} x_\alpha(u_\alpha)$. Denote by $\mathrm{Stab}_{T_n}(\bar{u})$ the stabilizer of $\bar{u}$ in T_n. Let $t \in T_n$. We may then note that

$$t \cdot \bar{u} = t \left(\prod_{\alpha \in S} x_\alpha(u_\alpha) \right) t^{-1} V_m = \prod_{\alpha \in S} t x_\alpha(u_\alpha) t^{-1} V_m = \prod_{\alpha \in S} x_\alpha(t^\alpha u_\alpha) V_m,$$

and hence

$$t \cdot \bar{u} = \bar{u} \iff \prod_{\alpha \in S} x_\alpha(t^\alpha u_\alpha) V_m = \prod_{\alpha \in S} x_\alpha(u_\alpha) V_m$$

$$\iff (t^\alpha - 1) u_\alpha \in \mathfrak{p}_L^m, \quad \text{for all } \alpha \in S.$$

It follows

$$\mathrm{Stab}_{T_n}(\bar{u}) = \{ t \in T_n \mid ord_L(t^\alpha - 1) \geq m - ord_L(u_\alpha), \text{ for all } \alpha \in S \}.$$

Of course, u_α's can be zero for some roots α. In this case $ord_L(u_\alpha) = \infty$ and the condition becomes vacuous. However, the requirement that $u_0 \notin V_n$ implies that for any $u \in u_0 V_n$ there will be at least one root α_0 such that $ord_L(u_{\alpha_0}) < n$.

For $m > n + ord_L(u_{\alpha_0})$, the condition $ord_L(t^{\alpha_0} - 1) \geq m - ord_L(u_{\alpha_0})$ determines a subgroup of T_n of index $q_L^{m - ord_L(u_{\alpha_0}) - n}$. Indeed α_0 is a homomorphism $T_n \to 1 + \mathfrak{p}_L^n$. The condition $ord_L(t^{\alpha_0} - 1) \geq m - ord_L(u_{\alpha_0})$ is equivalent to $t^{\alpha_0} \in 1 + \mathfrak{p}_L^{m - ord_L(u_{\alpha_0})}$. If $m - ord_L(u_{\alpha_0}) > n$, then $1 + \mathfrak{p}_L^{m - ord_L(u_{\alpha_0})}$ is a subgroup of index $q_L^{m - ord_L(u_{\alpha_0}) - n}$ in $1 + \mathfrak{p}_L^n$ and $\{ t \in T_n : t^{\alpha_0} \in 1 + \mathfrak{p}_L^{m - ord_L(u_{\alpha_0})} \}$ is a subgroup of the same index in T_n. The actual stabilizer $\mathrm{Stab}_{T_n}(\bar{u})$ is then a subgroup of this subgroup, and its index is a multiple of $q_L^{m - ord_L(u_{\alpha_0}) - n}$. Hence,

$$ord_p(|\mathrm{Orb}_{T_n}(\bar{u})|) \geq (m - ord_L(u_{\alpha_0}) - n) \, ord_p(q_L)$$

which tends to ∞ with m. $\square$

The action of T_0 on V_0 (by conjugation) induces the action of T_0 on $C(V_0, K)$ given by $t \cdot h(u) = h(t^{-1} u t)$. Then we define an action of T_0 on $K[[V_0]]$ by

$$\langle t \cdot \mu, h \rangle = \langle \mu, t^{-1} \cdot h \rangle \tag{8.1}$$

for $\mu \in K[[V_0]]$, $t \in T_0$, and $h \in C(V_0, K)$.

Lemma 8.9 *Let μ be a nonzero element of $K[[V_0]]$. Suppose there exists a character ξ of T_0 such that $t \cdot \mu = \xi(t)\mu$ for all $t \in T_0$. Then ξ is trivial and $\mu = c \cdot 1$ for some scalar c.*

Proof Take a nonzero $\mu \in K[[V_0]]$ and assume that $\langle \mu, t^{-1} \cdot h \rangle = \xi(t)\langle \mu, h \rangle$ for all $t \in T_0$.

Suppose first that ξ is not smooth and consider the characteristic function 1_{vV_n}, for some $v \in V_0$ and some $n \in \mathbb{N}$. Then there exists $t \in T_n$ such that $\xi(t) \neq 1$. Notice that $t^{-1} \cdot 1_{vV_n} = 1_{vV_n}$. This follows from Lemma 8.7 if $v \notin V_n$ and it holds trivially if $v \in V_n$. Then

$$\langle \mu, t^{-1} \cdot 1_{vV_n} \rangle = \langle \mu, 1_{vV_n} \rangle = \xi(t)\langle \mu, 1_{vV_n} \rangle$$

implies $\langle \mu, 1_{vV_n} \rangle = 0$. This condition forces $\langle \mu, h \rangle = 0$ for all smooth h, and then for all h. Thus we are reduced to the case when ξ is smooth.

To treat the case when ξ is smooth, we will show that $\langle \mu, 1_{u_0 V_n} \rangle = 0$ for any $u_0 \in V_0$ and positive integer n such that $u_0 V_n$ does not contain the identity.

There exists n_0 such that the restriction of ξ to T_{n_0} is trivial. Fix u_0 and n as above and assume $n \geq n_0$. For $m > n$,

$$1_{u_0 V_n} = \sum_{\bar{u} \in Q_m} 1_{\bar{u}}.$$

Hence

$$\langle \mu, 1_{u_0 V_n} \rangle = \sum_{\bar{u} \in Q_m} \langle \mu, 1_{\bar{u}} \rangle.$$

Now, let T_n act on V_0/V_m. We know from Lemma 8.7 that Q_m is preserved under this action. Write $[T_n \setminus Q_m]$ for a set of representatives of the distinct orbits in Q_m. Then

$$\langle \mu, 1_{u_0 V_n} \rangle = \sum_{\bar{u} \in [T_n \setminus Q_m]} \quad \sum_{t \cdot \bar{u} \in \mathrm{Orb}_{T_n}(\bar{u})} \langle \mu, t \cdot 1_{\bar{u}} \rangle$$

$$= \sum_{\bar{u} \in [T_n \setminus Q_m]} |\mathrm{Orb}_{T_n}(\bar{u})| \langle \mu, 1_{\bar{u}} \rangle$$

because $\langle \mu, t \cdot 1_{\bar{u}} \rangle = \xi(t^{-1})\langle \mu, 1_{\bar{u}} \rangle = \langle \mu, 1_{\bar{u}} \rangle$ for any $t \in T_n$. Since μ is a scalar multiple of some $\mu_0 \in o_K[[V_0]]$ and $\langle \mu_0, 1_{\bar{u}} \rangle \in o_K$, it follows that $\min_{\bar{u} \in Q_m} ord_p(\langle \mu, 1_{\bar{u}} \rangle)$ is bounded independently of m. By Lemma 8.8,

$$\lim_{m \to \infty} \min_{\bar{u} \in Q_m} ord_p(|\mathrm{Orb}_{T_n}(\bar{u})|) = \infty.$$

It follows that $\langle \mu, 1_{u_0 V_n} \rangle = 0$, for any $u_0 \in V_0$ and positive integer n such that $u_0 V_n$ does not contain the identity.

Thus μ is supported at the identity, so it is $c \cdot 1$ for some c. It now follows easily that ξ is trivial. $\qquad \square$

Corollary 8.10 *Let* $\xi : T_0 \to o_K^\times$ *be a continuous character. Let* $\iota : K \to K[[V_0]]$
be the natural embedding defined by $\iota(a) = a \cdot 1$*. Then*

$$\mathrm{Hom}_{K[[T_0]]}(K^{(\xi)}, K[[V_0]]) = \begin{cases} K \cdot \iota, & \text{if } \xi \text{ is trivial,} \\ 0, & \text{otherwise.} \end{cases}$$

Here, the action of $K[[T_0]]$ *on* $K[[V_0]]$ *is defined by equation* (8.1).

Proof Take $\psi \in \mathrm{Hom}_{K[[T_0]]}(K^{(\xi)}, K[[V_0]])$ and set $\mu = \psi(1)$. Then for any $t \in T_0$,

$$t \cdot \mu = \psi(t \cdot 1) = \xi(t)\psi(1) = \xi(t)\mu.$$

By Lemma 8.9, if $\xi \neq 1$, then $\mu = 0$, while for $\xi = 1$ we have $\mu = c \cdot 1$ for some
scalar c.   □

Recall from Sect. 7.2 that $K[[V_{w,\frac{1}{2}}^\pm]]^{(w\chi)}$ denotes the space $K[[V_{w,\frac{1}{2}}^\pm]]$ considered as a $K[[Q_{w,\frac{1}{2}}^\pm]]$-module, where $Q_{w,\frac{1}{2}}^\pm = T_0 V_{w,\frac{1}{2}}^\pm$, with the action of T_0 twisted
by the character $w\chi$ (see the discussion before Lemma 7.17). Similarly, we denote
by $K[[V_0]]^{(\chi_2)}$ the space $K[[V_0]]$ equipped with the action of $K[[T_0]]$ such that

$$\langle t \cdot \mu, h \rangle = \chi_2(t)\langle \mu, t^{-1} \cdot h \rangle \tag{8.2}$$

for $t \in T_0$, $\mu \in K[[V_0]]$, and $h \in C(V_0, K)$. As before, $t^{-1} \cdot h(u) = h(tut^{-1})$.

Corollary 8.11 *Suppose* $\chi_1, \chi_2 : T_0 \to o_K^\times$ *are continuous characters. Let* $\iota : K \to K[[V_0]]$ *be the natural embedding defined by* $\iota(a) = a \cdot 1$*. Then*

$$\mathrm{Hom}_{K[[T_0]]}(K^{(\chi_1)}, K[[V_0]]^{(\chi_2)}) = \begin{cases} K \cdot \iota, & \text{if } \chi_1 = \chi_2, \\ 0, & \text{otherwise.} \end{cases}$$

Proof The statement follows from Corollary 8.10, using the equality

$$\mathrm{Hom}_{K[[T_0]]}(K^{(\chi_1)}, K[[V_0]]^{(\chi_2)}) = \mathrm{Hom}_{K[[T_0]]}(K^{(\chi_1\chi_2^{-1})}, K[[V_0]]).$$

To prove the equality, write A for the action of T_0 on $K[[V_0]]$ given by Eq. (8.1)
and A_{χ_2} for the action of T_0 on $K[[V_0]]^{(\chi_2)}$ given by Eq. (8.2). Then $A_{\chi_2}(t).\mu =$
$\chi_2(t)A(t).\mu$. Hence, if $\psi \in \mathrm{Hom}_{K[[T_0]]}(K^{(\chi_1\chi_2^{-1})}, K[[V_0]])$ then

$$\chi_1\chi_2^{-1}(t)\psi(x) = \psi(\chi_1\chi_2^{-1}(t).x) = A(t).\psi(x),$$

whence

$$\chi_1(t)\psi(x) = \chi_2(t)A(t).\psi(x) = A_{\chi_2}(t).\psi(x).$$

So, ψ is also in $\mathrm{Hom}_{K[[T_0]]}(K^{(\chi_1)}, K[[V_0]]^{(\chi_2)})$. A similar argument shows containment in the other direction. □

8.2 Intertwining Algebra

Composing the embedding $\iota : K \to K[[G_0]]$ with the projection $K[[G_0]] \to M^{(\chi)}$, we obtain $\varphi : K \to M^{(\chi)}$ given by $\varphi(a) = (a \cdot 1) \otimes 1$. Since $K[[U_1^-]]^{(\chi)}$ embeds in $M^{(\chi)}$ as a direct summand and contains $1 \in G_0$, we see that φ is injective.

Theorem 8.12 *For any two continuous characters χ_1 and χ_2 of T_0, we have*

$$\mathrm{Hom}_{K[[P_0]]}(K^{(\chi_1)}, M^{(\chi_2)}) = \begin{cases} 0 & \text{if } \chi_1 \neq \chi_2, \\ K \cdot \varphi & \text{if } \chi_1 = \chi_2, \end{cases}$$

where $\varphi : K \to M^{(\chi_2)}$ sends $a \in K$ to $(a \cdot 1) \otimes 1$ in $M^{(\chi_2)}$.

Proof Since $M^{(\chi_2)} = \bigoplus_{v \in W} M_v^{(\chi_2)}$, as a $K[[B]]$-module and hence as a $K[[P_0]]$ module, we obtain

$$\mathrm{Hom}_{K[[P_0]]}(K^{(\chi_1)}, M^{(\chi_2)}) = \bigoplus_{v \in W} \mathrm{Hom}_{K[[P_0]]}(K^{(\chi_1)}, M_v^{(\chi_2)}).$$

We apply Proposition 8.4, taking w to be the identity element of W, which we denote e. Then $P_{\frac{1}{2}}^{w,\pm} = P_0$, and Proposition 8.4 implies $\mathrm{Hom}_{K[[P_0]]}(K^{(\chi_1)}, M_v^{(\chi_2)})$ $= 0$ for all $v \neq e$. Thus $\mathrm{Hom}_{K[[P_0]]}(K^{(\chi_1)}, M^{(\chi_2)}) = \mathrm{Hom}_{K[[P_0]]}(K^{(\chi_1)}, M_e^{(\chi_2)})$. Now, by Lemma 7.17, $M_e^{(\chi_2)}$ is isomorphic to $K[[U_1^-]]^{(\chi_2)}$, as a $K[[Q_{e,\frac{1}{2}}^\pm]]$-module and in particular as a $K[[T_0]]$-module. And, by Corollary 8.11,

$$\mathrm{Hom}_{K[[T_0]]}(K^{(\chi_1)}, K[[U_1^-]]^{(\chi_2)}) = \begin{cases} K \cdot \varphi, & \text{if } \chi_1 = \chi_2, \\ 0, & \text{otherwise.} \end{cases}$$

One readily confirms that φ is a $K[[P_0]]$-module map, completing the proof. □

Corollary 8.13 *For any two continuous characters χ_1 and χ_2 of P_0, we have*

$$\mathrm{Hom}_{K[[G_0]]}(M^{(\chi_1)}, M^{(\chi_2)}) = \begin{cases} 0 & \text{if } \chi_1 \neq \chi_2, \\ K \cdot \mathrm{id} & \text{if } \chi_1 = \chi_2. \end{cases}$$

Proof Similarly to the proof of Corollary 8.6, using adjoint associativity (Theorem 8.5) we have

$$\mathrm{Hom}_{K[[G_0]]}(M^{(\chi_1)}, M^{(\chi_2)}) = \mathrm{Hom}_{K[[G_0]]}(K[[G_0]] \otimes_{K[[P_0]]} K^{(\chi_1)}, M^{(\chi_2)})$$
$$\cong \mathrm{Hom}_{K[[P_0]]}(K^{(\chi_1)}, \mathrm{Hom}_{K[[G_0]]}(K[[G_0]], M^{(\chi_2)}))$$
$$\cong \mathrm{Hom}_{K[[P_0]]}(K^{(\chi_1)}, M^{(\chi_2)}).$$

The statement now follows from Theorem 8.12.   □

Since $M^{(\chi)}$ is the dual of $\mathrm{Ind}_{P_0}^{G_0}(\chi^{-1})$, we obtain the following result.

Corollary 8.14 *For any two continuous characters χ_1 and χ_2 of P_0, we have*

$$\mathrm{Hom}_{G_0}^c(\mathrm{Ind}_{P_0}^{G_0}(\chi_1), \mathrm{Ind}_{P_0}^{G_0}(\chi_2)) = \begin{cases} 0 & \text{if } \chi_1 \neq \chi_2, \\ K \cdot \mathrm{id} & \text{if } \chi_1 = \chi_2. \end{cases}$$

An analogous result for principal series representations of G was proved by Peter Schneider in an unpublished note. We deduce it from Corollary 8.14 as follows:

Proposition 8.15 *For any two continuous characters χ_1 and χ_2 of P, we have*

$$\mathrm{Hom}_G^c(\mathrm{Ind}_P^G(\chi_1), \mathrm{Ind}_P^G(\chi_2)) = \begin{cases} 0 & \text{if } \chi_1 \neq \chi_2, \\ K \cdot \mathrm{id} & \text{if } \chi_1 = \chi_2. \end{cases}$$

Proof If $\chi_1 = \chi_2$ or $\chi_1|_{P_0} \neq \chi_2|_{P_0}$, the statement follows immediately from Corollary 8.14.

It remains to consider the case when χ_1 and χ_2 are not equal but they have the same restriction to P_0. Set $\chi_0 = \chi_1|_{P_0} = \chi_2|_{P_0}$. Let

$$V_0 = \mathrm{Ind}_{P_0}^{G_0}(\chi_0^{-1}) \quad \text{and} \quad V_i = \mathrm{Ind}_P^G(\chi_i^{-1}), \ i = 1, 2.$$

Recall that

$$V_i = \{f : G \to K \text{ continuous} \mid f(gp) = \chi_i(p)f(g) \text{ for all } p \in P, g \in G\},$$

for $i = 1, 2$, and similarly for V_0. We know from Proposition 7.3 that the restriction $f \mapsto f|_{G_0}$ is a topological isomorphism from V_i to V_0, $i = 1, 2$. We prove that $\mathrm{Hom}_G^c(V_1, V_2) = 0$. Suppose, on the contrary, that there exists a nonzero intertwining operator $\varphi \in \mathrm{Hom}_G^c(V_1, V_2)$. By Corollary 8.14, $\varphi|_{V_0}$ is a nonzero scalar multiple of the identity operator, so by scaling we may assume $\varphi|_{V_0} = \mathrm{id}$. Take $f_1 \in V_1$ such that $f_1(1) = 1$ and let $f_0 = f_1|_{G_0}$. Select $t \in T$ such that $\chi_1(t) \neq \chi_2(t)$. Let $F_1 = L_t f_1$, the action by left translation. That is,

$F_1(g) = f_1(t^{-1}g)$, $g \in G$. Notice that

$$\varphi(F_1)(1) = F_1(1) = f_1(t^{-1}) = \chi_1(t^{-1}).$$

Next, let $f_2 = \varphi(f_1)$. This is the element of V_2 uniquely determined by the requirement $f_2|_{G_0} = f_0$. Since φ is an intertwining operator, we must have

$$\varphi(L_t f_1) = L_t f_2.$$

Evaluating the above functions at 1 gives us $\varphi(F_1)(1) = \chi_1(t^{-1})$ on the left and $f_2(t^{-1}) = \chi_2(t^{-1})$ on the right. This contradicts $\chi_1(t^{-1}) \neq \chi_2(t^{-1})$. □

Suppose $\chi : P \rightarrow K^\times$ is a continuous character, and set $\chi_0 = \chi|_{P_0}$. As discussed in Sect. 7.1, restriction to G_0 gives an isomorphism of Banach spaces $\mathrm{Ind}_P^G(\chi) \rightarrow \mathrm{Ind}_{P_0}^{G_0}(\chi_0)$. Not surprisingly, the G-representation $\mathrm{Ind}_P^G(\chi)$ differs significantly from the G_0-representation $\mathrm{Ind}_{P_0}^{G_0}(\chi_0)$. For example, we know from [76] that in the case of $\mathbf{G} = GL_2$, the $GL_2(\mathbb{Z}_p)$-representation $\mathrm{Ind}_{P_0}^{G_0}(\chi_0)$ can have infinitely many finite dimensional subrepresentations, while the $GL_2(\mathbb{Q}_p)$-representation $\mathrm{Ind}_P^G(\chi)$, if reducible, has a unique irreducible subrepresentation. With this in mind, the result of Corollary 8.14 for G_0 seems surprising. Examples of $\mathrm{Ind}_{P_0}^{G_0}(\chi_0)$ with infinitely many finite dimensional subrepresentations for a general group $\mathbf{G}$ are constructed in Sect. 8.3.

8.2.1 Ordinary Representations of $GL_2(\mathbb{Q}_p)$

Let $G = GL_2(\mathbb{Q}_p)$. As before, T is the group of diagonal matrices and $P = TU$ is the group of upper triangular matrices. Recall that an irreducible admissible K-Banach space representation of $GL_2(\mathbb{Q}_p)$ is called ordinary if it is a subquotient of a continuous principal series induced from a unitary character.

Proposition 8.16 *Ordinary representations of $G = GL_2(\mathbb{Q}_p)$ are*

(i) $\chi \circ \det$ and $\chi \circ \det \otimes \widehat{\mathrm{St}}$, where χ is a unitary character of $\mathbb{Q}_p^\times$, and

(ii) $\mathrm{Ind}_P^G(\chi_1 \otimes \chi_2)$, where χ_1, χ_2 are unitary characters of $\mathbb{Q}_p^\times$ such that $\chi_1 \neq \chi_2$.

All these representations are pairwise inequivalent.

Proof The irreducible components of all principal series induced from unitary characters are listed in Propositions 7.10. We have to show that they are pairwise inequivalent. This clearly holds for the representations listed under (i). In addition, Proposition 8.15 tells us that the representations listed under (ii) are pairwise inequivalent. Finally, if $\chi_1 \neq \chi_2$ then $\mathrm{Ind}_P^G(\chi_1 \otimes \chi_2)$ is not isomorphic to $\chi \circ \det \otimes \widehat{\mathrm{St}}$

for any χ. Otherwise, the exact sequence

$$0 \to \chi \circ \det \to \mathrm{Ind}_P^G(\chi \otimes \chi) \to \chi \circ \det \otimes \widehat{\mathrm{St}} \to 0$$

would give us a nonzero intertwining operator $\mathrm{Ind}_P^G(\chi \otimes \chi) \to \mathrm{Ind}_P^G(\chi_1 \otimes \chi_2)$, thus contradicting Proposition 8.15. $\qquad\square$

8.3   Finite Dimensional G_0-Invariant Subspaces

In this section we discuss finite dimensional G_0-invariant subspaces of the representation $V = \mathrm{Ind}_P^G(\chi^{-1})$ in the case when $\chi : T \to K^\times$ is locally algebraic. This means that

$$\chi = \chi_{\mathrm{alg}} \chi_{\mathrm{sm}},$$

with χ_{sm} smooth and χ_{alg} L-algebraic. A similar, but more general discussion, with χ_{alg} replaced by a $\mathbb{Q}_p$-rational character, can be found in Appendix to [4].

Recall that we introduced several types of parabolic induction. Then, we can consider the smooth induction by χ_{sm} and the algebraic induction by χ_{alg}. Set $U = \mathrm{Ind}_P^G(\chi_{\mathrm{sm}}^{-1})^{\mathrm{sm}}$. Suppose in addition that χ_{alg} is dominant. Then we have a nonzero algebraically induced representation $W = \mathrm{ind}_P^G(\chi_{\mathrm{alg}}^{-1})$, which is finite dimensional and irreducible. Given $f \in U$ and $h \in W$, let fh be the pointwise product, $(fh)(x) = f(x)h(x)$. We see immediately that the product $fh : G \to K$ is continuous and satisfies

$$(fh)(gp) = f(gp)h(gp) = \chi_{\mathrm{sm}}(p)f(g)\chi_{\mathrm{alg}}(p)h(g) = \chi(p)(fh)(g)$$

for all $p \in P$ and $g \in G$, so $fh \in V$. The tensor product $U \otimes_K W$ is a locally algebraic representation (see Appendix to [68] or Section 4.2 in [32] for the definition and basic properties of such representations). Pointwise multiplication of functions gives us a natural map

$$U \otimes_K W \to V.$$

This map is injective, which follows from [51], using exactness of the functor $\mathcal{F}_P^G$ for the split group $\mathbf{G}$.

Now, we consider the corresponding G_0-representations. The algebraic representation W remains irreducible when restricted to G_0. By Proposition 6.21, U decomposes as a countable direct sum of finite dimensional representations ρ with finite multiplicities

$$U \cong \bigoplus_{\rho} m(\rho)\rho.$$

Then V contains

$$U \otimes_K W \cong \bigoplus_{\rho} m(\rho)(\rho \otimes_K W). \tag{8.3}$$

Note that every subspace $\rho \otimes_K W$ is finite-dimensional, and hence closed in V. Alternatively, we can use Corollary 4.2.9 of [32] to show that $U \otimes_K W$ decomposes as a direct sum of irreducible finite-dimensional representations. In conclusion, the G_0-representation V contains countably many finite-dimensional topologically irreducible subrepresentations. Still, by Lemma 8.14, $\operatorname{Hom}_{G_0}(V, V) = K \cdot \operatorname{id}$.

On the other hand, the decomposition (8.3) implies that $U \otimes_K W$, as a G_0-representation, has numerous self-intertwining operators that are not scalar multiples of the identity. These operators, however, cannot be extended continuously to V. Here is an example by Matthias Strauch.

8.3.1   Induction from the Trivial Character: Intertwiners

Let $G = GL_2(\mathbb{Q}_p)$ and $G_0 = GL_2(\mathbb{Z}_p)$. Let $P = TU$ be the group of upper triangular matrices, with T the group of diagonal matrices in G and U the unipotent radical of P. We consider the principal series $V = \operatorname{Ind}_P^G(\mathbf{1})$. Let $W = \operatorname{Ind}_P^G(\mathbf{1})^{\mathrm{sm}}$, which is a dense subspace of V. As discussed on page 155, we have the following exact sequences of representations of G

$$0 \to \mathbf{1} \to W \to \mathrm{St} \to 0 \quad \text{and} \quad 0 \to \mathbf{1} \to V \to \widehat{\mathrm{St}} \to 0.$$

The trivial representation $\mathbf{1}$ is realized on the one-dimensional subspace $W_1 \subset W$ consisting of all constant functions $G_0 \to K$. To find its complement in W, we fix the left Haar integral on G_0 with values in K such that $\int_{G_0} dx = 1$. Define

$$W_2 = \left\{ f \in W \mid \int_{G_0} f(x)dx = 0 \right\}.$$

This is a G_0-invariant subspace of W, and $W = W_1 \oplus W_2$. Define $\varphi : W \to W$ by

$$\varphi(f) = \int_{G_0} f(x)dx,$$

where the number $\varphi(f) \in K$ is interpreted as the constant function $G_0 \to K$ with the value $\varphi(f)$. Then φ is the projection onto W_1 and also an intertwining operator. Hence, φ is an intertwining operator on W which is not a scalar multiple of id_W.

Next, we consider W equipped with the sup norm (the norm inherited from V). We will show that φ is not bounded, and hence it is not continuous. For that, we take the following decomposition (see Proposition 5.45)

$$G_0 = U_1^- P_0 \coprod \dot{w} U_0^- P_0,$$

where $\dot{w} = \begin{pmatrix} 0 & 1 \\ 1 & 0 \end{pmatrix}$. Then, $W \cong C^\infty(U_1^-, K) \oplus C^\infty(U_0^-, K)$. We identify U_0^- with $\mathbb{Z}_p$. For $n \in \mathbb{N}$, define $h \in C^\infty(U_0^-, K)$ by

$$h_n(x) = \begin{cases} 1, & x \in p^n \mathbb{Z}_p \\ 0, & \text{otherwise.} \end{cases}$$

Let f_n be the function in W corresponding to h_n. It is given by

$$f_n(x) = \begin{cases} 1, & x \in \dot{w} U_n^- P_0 \\ 0, & \text{otherwise.} \end{cases}$$

Then $\| f_n \| = \sup_{x \in G_0} |f_n(x)|_K = 1$. On the other hand, let $c = \int_{\dot{w} U_0^- P_0} dx$. Then $\int_{\dot{w} U_n^- P_0} dx = cp^{-n}$ and

$$\varphi(f_n) = \int_{G_0} f(x) dx = \int_{\dot{w} U_n^- P_0} dx = cp^{-n}.$$

The norm $|\ |_K$ satisfies $|p|_K = p^{-d}$, where d is the degree of the field extension $K/\mathbb{Q}_p$ (see Appendix A.2.2). It follows that

$$\frac{\|\varphi(f_n)\|}{\| f_n \|} = |c|_K p^{dn},$$

proving that φ is unbounded. As such, it cannot be extended continuously to V.

Exercise 8.17 With notation as above, prove that W_2 is dense in V.

8.4 Reducibility of Principal Series

One of the basic questions in the representation theory is to decide whether an induced representation is reducible. Among smooth representations, there is a nice and important class called discrete series representations. For them, the reducibility of a parabolically induced representation is completely determined by intertwining operators. They define the R-group, which governs not only reducibility, but also the decomposition of the induced space.

As we see from Proposition 8.15, we cannot hope for anything similar for principal series representations on p-adic Banach spaces. Away from scalar multiples of the identity map, there are no intertwiners. So, we need other methods for determining reducibility.

In this section, we give an overview of some reducibility results. The case $L = \mathbb{Q}_p$ is best understood.

8.4.1 Locally Analytic Vectors

So far, locally analytic representations were mentioned in passing but we didn't use them in proofs. Now, to tell more about the reducibility of continuous principal series, we will tap into the knowledge on locally analytic representations. And this knowledge is extensive, so we will just mention what we need.

Take a continuous character $\chi : T \to K^{\times}$. Then χ is locally $\mathbb{Q}_p$-analytic, and the induced representation $\mathrm{Ind}_P^G(\chi^{-1})$ contains the dense subset $\mathrm{Ind}_P^G(\chi^{-1})^{\mathbb{Q}_p-\mathrm{an}}$ of locally $\mathbb{Q}_p$-analytic vectors. As in [50], there is a canonically associated Verma module, and whenever it is simple, $\mathrm{Ind}_P^G(\chi^{-1})^{\mathbb{Q}_p-\mathrm{an}}$ is topologically irreducible. This was proved by Frommer in [33] for $L = \mathbb{Q}_p$ and G split. The general case, with L a finite extension of $\mathbb{Q}_p$ and G a connected reductive algebraic group over L, was proved by Orlik and Strauch in [50].

Reducibility Question for $\mathbf{G}(\mathbb{Q}_p)$

The theory we developed so far was for $\mathbf{G}(L)$, where $\mathbb{Q}_p \subseteq L \subseteq K$ is a sequence of finite extensions. Our methods worked equally well for an arbitrary L as for $L = \mathbb{Q}_p$. However, there are parts of the theory of Banach space representations that work only for $\mathbf{G}(\mathbb{Q}_p)$.

Assume in this section that $L = \mathbb{Q}_p$ and $G = \mathbf{G}(\mathbb{Q}_p)$. Then, as mentioned in Sect. 4.4.1, if V is an admissible Banach space representation of G, the space of locally analytic vectors $V^{\mathrm{an}} = V^{\mathbb{Q}_p-\mathrm{an}}$ is dense in V. Moreover, the functor $V \mapsto V^{\mathrm{an}}$ is exact. We apply this to $V = \mathrm{Ind}_P^G(\chi^{-1})$. Then, V^{an} is the locally analytic principal series whose reducibility is well-understood. By the exactness of $V \mapsto V^{\mathrm{an}}$ and the density of V^{an} in V, if V^{an} is irreducible, then V is irreducible as well. However, if V^{an} reduces, V may be either reducible or irreducible.

Assume that $\mathbf{G}$ is semisimple and simply connected. Let $\lambda_1, \ldots, \lambda_r$ be the fundamental weights and $\delta = \sum_{i=1}^{r} \lambda_i$. The assumption that $\mathbf{G}$ is simply connected implies that $\lambda_i \in X(\mathbf{T})$, for all i, and so $\delta \in X(\mathbf{T})$. As explained on page 111, we use the same symbol δ for the corresponding character $\delta : T \to \mathbb{Q}_p^\times$. Similarly, for a root α, we have the coroot $\alpha^\vee : \mathbb{Q}_p^\times \to T$.

The character $\chi : T \to K^\times$ is called **anti-dominant** if

$$\chi \delta \circ \alpha^\vee \neq (\ )^m \tag{8.4}$$

for any integer $m \geq 1$ and any positive root α. In [61, Conjecture 2.5], Schneider conjectures that the G-representation $\mathrm{Ind}_P^G(\chi^{-1})$ is topologically irreducible if χ is anti-dominant. It is explained there that the conjecture holds for $GL_2(\mathbb{Q}_p)$. (The group $GL_2(\mathbb{Q}_p)$ is not semisimple, but the anti-dominance condition is well-defined.)

Example 8.18 Let $G = GL_2(\mathbb{Q}_p)$. As before, $P = TU$ is the group of upper triangular matrices. Let $\delta_P : T \to K^\times$ be the modulus character of P, $\delta_P(\mathrm{diag}(a,b)) = |ab^{-1}|_p$. Then $(\delta_P^{-1} \circ \alpha^\vee)(a) = |a^{-2}|_p$, which is not an algebraic character. It follows that δ_P^{-1} is anti-dominant, and $\mathrm{Ind}_P^G(\delta_P)$ is irreducible.

The proposition below describes not only G-irreducibility, but also the much stronger G_0-irreducibility.

Proposition 8.19 *Suppose* **G** *is semisimple and simply connected, and* $G = $ **G**$(\mathbb{Q}_p)$. *If the anti-dominance condition for* χ *continues to hold after restriction to an arbitrary small open subgroup of* $\mathbb{Q}_p^\times$ *then*

(i) $M^{(\chi)}$ *is simple as a* $K[[G_0]]$-module, and
(ii) $\mathrm{Ind}_{P_0}^{G_0}(\chi^{-1})$ *is topologically irreducible as a* G_0-representation.

Proof This is Proposition 2.6 (ii) from [61]. Notice that (i) and (ii) are equivalent by duality (Corollary 4.46). The proof follows from the corresponding facts for the locally analytic principal series and the density of analytic vectors in admissible Banach space representations. More details can be found in [61].   □

An example of a character not satisfying the anti-dominance condition is what else but a dominant character.

Example 8.20 Suppose $\chi : T \to K^\times$ is a dominant algebraic character. Take a positive root α and set $n = \langle \chi, \alpha^\vee \rangle$. Then n is a non-negative integer and we have $\chi(\alpha^\vee(a)) = a^n$, for all $a \in \mathbb{Q}_p^\times$. In addition, $\langle \delta, \alpha^\vee \rangle \geq 1$. As we discussed in Sect. 7.1.3, the representation $\mathrm{Ind}_P^G(\chi^{-1})$ is reducible. It contains the algebraic representation $\mathrm{ind}_P^G(\chi^{-1})$.

Reducibility Question for G(L)

Let $L \subseteq K$ be an arbitrary finite extension of $\mathbb{Q}_p$.

Example 8.21 As above, if $\chi : T \rightarrow K^\times$ is a dominant algebraic character, then $\mathrm{ind}_P^G(\chi^{-1})$ is a closed finite-dimensional G-invariant subspace of $\mathrm{Ind}_P^G(\chi^{-1})$. Let σ be an automorphism of K which is nontrivial on L. Given $f \in \mathrm{ind}_P^G(\chi^{-1})$, the function $F = \sigma \circ f$ satisfies

$$F(gp) = \sigma(f(gp)) = (\sigma \circ \chi)(p)F(g),$$

for all $g \in G$, $p \in P$. It follows that $\{\sigma \circ f \mid f \in \mathrm{ind}_P^G(\chi^{-1})\}$ is a closed finite-dimensional G-invariant subspace of $\mathrm{Ind}_P^G((\sigma \circ \chi)^{-1})$ and therefore $\mathrm{Ind}_P^G((\sigma \circ \chi)^{-1})$ is reducible. On the other hand, if $\chi \neq 1$, then $\sigma \circ \chi$ satisfies the condition (8.4) because on the left hand side we have a non-algebraic character, and it cannot be equal to the algebraic character $(\ )^m$.

Thus, the anti-dominance condition (8.4) does not predict irreducibility if $L \neq \mathbb{Q}_p$. To understand the reason, we have to consider G as a locally $\mathbb{Q}_p$-analytic group. This is done by the restriction of scalars $\mathrm{Res}_{L/\mathbb{Q}_p}$. The group G can be identified with the group of $\mathbb{Q}_p$-points of $\mathrm{Res}_{L/\mathbb{Q}_p} \mathbf{G}$, which is an algebraic group defined over $\mathbb{Q}_p$. However, this group is not split. For non-split groups, the anti-dominance condition (8.4) has to be modified.

The following example shows how we can use locally $\mathbb{Q}_p$-analytic vectors to get the irreducibility of continuous principal series.

Example 8.22 We consider $G = GL_2(L)$ as in [50, Example 4.2.2].

Let Γ denote the set of $\mathbb{Q}_p$-embeddings of L into K. Suppose that the cardinality of Γ is equal to $[K : L]$. Let T be the set of diagonal matrices, and P the set of upper triangular matrices. Let $\chi : T \rightarrow K^\times$ be a continuous character. There are $c_{1,\sigma}, c_{2,\sigma} \in K$ such that, for $t_1, t_2 \in L^\times$ sufficiently close to 1, we can write

$$\chi \begin{pmatrix} t_1 & 0 \\ 0 & t_2 \end{pmatrix} = \prod_{\sigma \in \Gamma} \sigma(t_1)^{c_{1,\sigma}} \sigma(t_2)^{c_{2,\sigma}}.$$

If for all $\sigma \in \Gamma$ we have $(c_{1,\sigma} - c_{2,\sigma}) \notin \mathbb{Z}_{\geq 0}$, then the locally $\mathbb{Q}_p$-analytic principal series is irreducible. This implies that $\mathrm{Ind}_P^G(\chi^{-1})$ is irreducible.

8.4.2 A Criterion for Irreducibility

Back to our general case, with $\mathbf{G}$ a split connected reductive $\mathbb{Z}$-group and $G = \mathbf{G}(L)$, we will present an irreducibility criterion from [3]. First, we note that by Lemma 6.5, there is a basis $\lambda_1, \ldots, \lambda_r$ for $X(\mathbf{T})$ consisting of dominant elements.

Let $\eta : L^\times \rightarrow K^\times$ be a continuous character. As η must map $o_L^\times$ into $o_K^\times$, it induces a map $L^\times/o_L^\times \rightarrow K^\times/o_K^\times$. That is, for $a \in L^\times$, the valuation $v_K(\eta(a))$

depends only on $v_L(a)$. Let $e(\eta)$ denote the integer such that

$$v_K \circ \eta = e(\eta) \cdot v_L. \tag{8.5}$$

Theorem 8.23 *Let* $\chi_1, \ldots, \chi_r : L^\times \to K^\times$ *be continuous characters such that* $e(\chi_i) < 0$ *for* $1 \le i \le r$. *Define* $\chi : T \to K^\times$ *by* $\chi(t) = \prod_{i=1}^{r} \chi_i(t^{\lambda_i})$. *Then* $\mathrm{Ind}_P^G \chi^{-1}$ *is topologically irreducible.*

Notice that we can check whether $e(\chi_i) < 0$ simply by evaluating $\chi_i(\varpi_L)$; no information about χ_i near 1 is needed. On the other hand, the irreducibility results from the previous section are given in terms of the behavior of χ_i in a small neighborhood of 1. This is because they use locally analytic vectors (see Proposition 8.19 and Example 8.22). In any case, Proposition 8.19 and Theorem 8.23 give irreducibility for different sets of characters.

For the proof of Theorem 8.23, we use the duality and prove the corresponding result for $M^{(\chi)}$ (Theorem 8.27). As discussed in Sect. 7.2.1, $M^{(\chi)}$ is isomorphic to a quotient of $K[[G_0]]$. For $\mu \in K[[G_0]]$, we denote by $[\mu]$ its image in $M^{(\chi)}$, that is,

$$[\mu] = \mu \otimes 1 \in M^{(\chi)}.$$

By Proposition 7.14, any $[\mu] \in M^{(\chi)}$ has a representative of the form

$$\eta = \sum_{w \in W} \dot{w} \eta_w, \quad \eta_w \in K[[U^-_{w,\frac{1}{2}}]].$$

The idea of the proof of Theorem 8.27 can be explained by the following simple example.

Example 8.24 Let $G = GL_2(L)$, with $P = TU$ the group of upper triangular matrices. Notice that $U^-_{w,\frac{1}{2}} = U_0^-$. Let

$$t = \begin{pmatrix} a & 0 \\ 0 & b \end{pmatrix} \in T, \ \dot{w} = \begin{pmatrix} 0 & 1 \\ 1 & 0 \end{pmatrix}, \ u = \begin{pmatrix} 1 & 0 \\ x & 1 \end{pmatrix} \in U_0^-, \ v = \begin{pmatrix} 1 & 0 \\ ab^{-1}x & 1 \end{pmatrix}, \ \text{and } s = \begin{pmatrix} b & 0 \\ 0 & a \end{pmatrix}.$$

Take $v = \dot{w}u$. To find the action of t on $[v]$, we use the following formula

$$\begin{pmatrix} a & 0 \\ 0 & b \end{pmatrix} \begin{pmatrix} 0 & 1 \\ 1 & 0 \end{pmatrix} \begin{pmatrix} 1 & 0 \\ x & 1 \end{pmatrix} = \begin{pmatrix} 0 & 1 \\ 1 & 0 \end{pmatrix} \begin{pmatrix} 1 & 0 \\ ab^{-1}x & 1 \end{pmatrix} \begin{pmatrix} b & 0 \\ 0 & a \end{pmatrix}$$

Suppose that N is a $(K[[G_0]], G)$-submodule of $M^{(\chi)}$ which contains $[1 + v]$. If $ab^{-1}x \in o_L$, then

$$\chi(t^{-1})t \cdot [1 + v] = [1 + \chi(t^{-1}s)\eta],$$

where $\eta = \dot{w}v \in \dot{w}o_K[[U_0^-]]$. If we impose the additional condition that $\chi(t^{-1}s) \in \mathfrak{p}_K$, then

$$1 + \chi(t^{-1}s)\eta \in 1 + \varpi_K \cdot o_K[[G_0]].$$

Such an element is invertible in $o_K[[G_0]]$, by Exercise 2.48. Consequently, N contains [1], and so $N = M^{(\chi)}$.

We can apply a similar method in the general case. Let N be a nonzero $(K[[G_0]], G)$-submodule of $M^{(\chi)}$. The requirement that N contains $[1 + \dot{w}u]$ is, of course, very strong, but it is not needed. It turns out that N always contains an element of the form $[1 + v]$, where v satisfies the following two conditions:

1. $v \in o_K[[G_0]]$, and
2. the support of v is disjoint from $G_n P_0$, for some n.

Such $[1 + v]$ serves us well: with an appropriate choice of $t \in T$ and an appropriate condition on χ, we can repeat the same trick as above to show that $N = M^{(\chi)}$.

We will work with the image of $o_K[[G_0]]$ in $M^{(\chi)}$, which is equal to $M_0^{(\chi)} = o_K[[G_0]] \otimes_{o_K[[P_0]]} o_K^{(\chi)}$. We need the following two technical lemmas from [3].

Lemma 8.25 *Fix $n \in \mathbb{N}$. With assumptions as in Theorem 8.23, there exists $t \in T$ such that*

$$\chi(t^{-1})t \cdot v \in [\varpi_K \cdot o_K[[G_0]]]$$

for any $v \in M_0^{(\chi)}$ that vanishes on $G_n P$.

Proof This is Corollary 7.4 in [3]. It follows from Lemma 5.1 in [3], the main technical result of that paper, which describes the action of T on a convenient, explicit model for the space G/P. $\qquad\square$

Lemma 8.26 *Fix $w_0 \in W$ and $u_0 \in U_{w_0, \frac{1}{2}}^-$. Let $n \geq 1$. Then*

$$u_0^{-1}\dot{w}_0^{-1} \cdot (\coprod_w \dot{w}U_{w, \frac{1}{2}}^-) \cap G_n P_0 = U_n^-,$$

that is, if $w \in W$, $u \in U_{w, \frac{1}{2}}^-$ and $u_0^{-1}\dot{w}_0^{-1}\dot{w}u \in G_n P_0$, then $w = w_0$ and $u_0^{-1}u \in U_n^-$.

Proof Consider the projection from G_0 to $\bar{G}$ (the points of $\mathbf{G}$ over the finite field $l = o_L/\mathfrak{p}_L$). The sets $\dot{w}U_{w, \frac{1}{2}}^-$ for $w \in W$ all project to distinct Bruhat cells. Hence

$$w_0 u_0 G_1 P_0 \cap \dot{w}U_{w, \frac{1}{2}}^- \neq \emptyset \implies w = w_0.$$

Assume $w = w_0$. Then $u_0^{-1} \dot{w}_0^{-1} \dot{w} U_{w,\frac{1}{2}}^{-} = u_0^{-1} U_{w,\frac{1}{2}}^{-} \subset U_0^{-}$. Hence, it is enough to prove $U_0^{-} \cap G_n P_0 \subset G_n$. To show this, we consider the projection to G_0/G_n. Since $U_0^{-} \cap P_0 = \{1\}$, the only element of G_0/G_n which is in the image of both P_0 and U_0^{-} is the identity. $\qquad\square$

The following is Theorem 7.5 from [3] and its proof.

Theorem 8.27 *Define* $\chi : T \to K^{\times}$ *by* $\chi(t) = \prod_{i=1}^{r} \chi_i(t^{\lambda_i})$ *where* $\lambda_1, \ldots \lambda_r$ *are dominant and form a basis for* $X(\mathbf{T})$. *Assume that* $e(\chi_i) < 0$ *for* $1 \leq i \leq r$. *Then* $M^{(\chi)}$ *has no proper nontrivial G-invariant* $K[[G_0]]$*-submodules.*

Proof Choose a nontrivial element of $M^{(\chi)}$, and construct a representative $\eta = \sum_{w \in W} \dot{w} \eta_w$ with $\eta_w \in K[[U_{w,\frac{1}{2}}^{-}]]$ for each $w \in W$. Let N be the $(K[[G_0]], G)$-submodule generated by $[\eta]$. We want to show that $N = M^{(\chi)}$.

By scaling, we may assume that $\eta_w = (\eta_{w,\ell})_{\ell=0}^{\infty} \in o_K[[U_{w,\frac{1}{2}}^{-}]]$ for each w, and that there exists $n \geq 1$, $w_0 \in W$ and $\bar{u}_0 \in U_{w_0,\frac{1}{2}}^{-}/U_n^{-}$ such that the coefficient c_0 of $\bar{u}_0$ in $\eta_{w_0,n}$ is a unit. Choose $u_0 \in U_{w_0,\frac{1}{2}}^{-}$ which projects to $\bar{u}_0$, and let $\mu = u_0^{-1} \dot{w}_0^{-1} \eta$. If we write μ as an element of the projective limit $\mu = (\mu_\ell)_{\ell=0}^{\infty}$, then

$$\mu_n = c_0 + \sum_{\substack{\bar{g} \in G_0/G_n \\ \bar{g} \neq 1}} c_{\bar{g}} \bar{g}, \qquad c_0 \in o_K^{\times}, c_{\bar{g}} \in o_K. \tag{8.6}$$

Now, observe that the partition of G_0 as $G_n \cup (G_0 \smallsetminus G_n)$ gives rise to direct sum decompositions of $C(G_0, K), o_K[[G_0]]$ and $K[[G_0]]$. Moreover $\{\lambda \in K[[G_0]] : \text{supp}(\lambda) \subset G_n\}$ is canonically identified with $K[[G_n]]$. Let

$$\mu = \mu' + \mu'', \qquad \mu' \in o_K[[G_n]], \text{ supp}(\mu'') \subset G_0 \smallsetminus G_n,$$

and note that, by Lemma 8.26, the support of μ'' is actually disjoint from $G_n P_0$. The image of μ' under the augmentation map is precisely c_0, the coefficient of the identity coset of μ_n in Eq. (8.6). Since c_0 is a unit, we know from Proposition 5.42 that μ' is an invertible element of $o_K[[G_n]]$. Multiplying by its inverse, we obtain an element of the form $1 + \nu$ where the support of ν is disjoint from $G_n P_0$.

Note that the elements $[\eta]$, $[\mu]$ and $[1 + \nu]$ generate the same submodule N of $M^{(\chi)}$. Now, choose t as in Lemma 8.25. If we act by $\chi(t)^{-1} t$ on $[1 + \nu]$, we see that N contains

$$\chi(t)^{-1} t \cdot [1 + \nu] = [1 + \chi(t)^{-1} t \cdot \nu] \in [1 + \varpi_K \cdot o_K[[G_0]]].$$

From Exercise 2.48 or [3, Lemma 6.1], we know that the elements of $1 + \varpi_K \cdot o_K[[G_0]]$ are units. Hence, $N = M^{(\chi)}$. $\qquad\square$

Finally, Theorem 8.23 follows from Theorem 8.27, using the duality between $\text{Ind}_P^G \chi^{-1}$ and $M^{(\chi)}$.

Appendix A
Nonarchimedean Fields and Spaces

A.1 Ultrametric Spaces

A metric space (X, d) is called an ultrametric space if d satisfies the strong triangle inequality (see Definition A.1). Many of the objects we consider in this book are ultrametric: the field of p-adic numbers $\mathbb{Q}_p$, any finite extension $K/\mathbb{Q}_p$, as well as any normed K-vector space.

An ultrametric space has staggering topological properties that could look strange to the novices to the p-adic world. In this section, we prove some of these properties.

Definition A.1

(i) Let X be a nonempty set. A **metric**, or distance, on X is a function $d : X \times X \to \mathbb{R}_{\geq 0}$ such that

    (1) $d(x, y) = 0$ if and only if $x = y$,
    (2) $d(x, y) = d(y, x)$,
    (3) $d(x, y) \leq d(x, z) + d(z, y)$ (the triangle inequality),

    for all $x, y, z \in X$. A set X together with a metric d is called a **metric space**.

(ii) A metric d on a set X is called an **ultrametric** or a **nonarchimedean metric** if

    (3') $d(x, y) \leq max(d(x, z), d(z, y))$ (the strong triangle inequality)

    for all $x, y, z \in X$.

(iii) An **ultrametric space** is a pair (X, d) consisting of a set X together with an ultrametric d on X.

© The Author(s), under exclusive license to Springer Nature Switzerland AG 2022
D. Ban, *p-adic Banach Space Representations*, Lecture Notes
in Mathematics 2325, https://doi.org/10.1007/978-3-031-22684-7

Proposition A.2 *(The Isosceles Triangle Principle) Let (X, d) be an ultrametric space. Let $x, y, z \in X$. If $d(x, y) \neq d(y, z)$, then*

$$d(x, z) = \max\{d(x, y), d(y, z)\}.$$

Hence, every "triangle" is isosceles, and if it is not equilateral, then the legs are longer than the base.

Proof We may assume without loss of generality that $d(x, y) < d(y, z)$. Notice that $d(y, z) \leq \max\{d(y, x), d(x, z)\}$ and this maximum must be equal to $d(x, z)$ because it is strictly greater than $d(x, y)$. It follows

$$d(y, z) \leq d(x, z) \leq \max\{d(x, y), d(y, z)\} = d(y, z),$$

which implies $d(x, z) = d(y, z)$.   □

Let (X, d) be a metric space. Then for $a \in X$ and a real number $r > 0$, the **closed ball** (respectively **open ball**) of radius r centered at a is

$$B_r(a) = \{x \in X \mid d(x, a) \leq r\}, \quad \text{respectively,} \quad B_r^-(a) = \{x \in X \mid d(x, a) < r\}.$$

The open balls form the base for a topology on X, making it a topological space. The fundamental property of ultrametric spaces is that all balls are both open and closed.

Proposition A.3 *Let (X, d) be an ultrametric space. Let $a \in X$ and $r > 0$.*

   (i) *Every point inside a ball is its center. More specifically, if $b \in B_r(a)$, then $B_r(a) = B_r(b)$ and if $b \in B_r^-(a)$, then $B_r^-(a) = B_r^-(b)$.*
  (ii) *Intersecting balls are contained in each other. That is, if $B_r(a) \cap B_s(b) \neq \emptyset$, then either $B_r(a) \subseteq B_s(b)$ or $B_s(b) \subseteq B_r(a)$.*
 (iii) *$B_r(a)$ is both open and closed in X. Similarly, $B_r^-(a)$ is both open and closed in X.*

Proof

  (i) Let $b \in B_r(a)$. Then

$$x \in B_r(a) \implies d(x, a) \leq r$$
$$\implies d(x, b) \leq max(d(x, a), d(a, b)) \leq r$$
$$\implies x \in B_r(b).$$

It follows $B_r(a) \subseteq B_r(b)$. In the same way, $B_r(b) \subseteq B_r(a)$. Therefore, $B_r(a) = B_r(b)$. The proof for $B_r^-(a)$ is analogous.

(ii) Suppose $B_r(a) \cap B_s(b) \neq \emptyset$ and take $x \in B_r(a) \cap B_s(b)$. By (i), $B_r(a) = B_r(x)$ and $B_s(x) = B_s(b)$. If $r \leq s$, then

$$B_r(a) = B_r(x) \subseteq B_s(x) = B_s(b).$$

Otherwise, we show in the same way that $B_s(b) \subset B_r(a)$.

(iii) If $x \in B_r(a)$, then $B_r^-(x) \subseteq B_r(x) = B_r(a)$. This shows that $B_r(a)$ is open.

To prove that $B_r^-(a)$ is closed, we will show that its complement $C = X \setminus B_r^-(a)$ is open. If $C = \emptyset$, then C is open. Otherwise, take $x \in C$. Since $x \notin B_r^-(a)$, it follows $d(x, a) \geq r$. It is easy to show that the balls $B_r^-(x)$ and $B_r^-(a)$ are disjoint. It follows that $B_r^-(x) \subseteq C$ is an open neighborhood of x in C. This shows that C is open.

$\square$

Corollary A.4 *Let (X, d) be an ultrametric space and $r > 0$. Then X is a disjoint union of open balls of radius r. Similarly, X is a disjoint union of closed balls of radius r.*

As a consequence, there are many locally constant functions on an ultrametric space. Another consequence is that ultrametric spaces are totally disconnected.

Recall that a nonempty topological space X is **totally disconnected** if the only connected components of X are the singletons.

Proposition A.5 *Let (X, d) be an ultrametric space. Then X is totally disconnected.*

Proof Exercise. $\square$

A.2   Nonarchimedean Local Fields

In this section, we give an overview of the properties of nonarchimedean local fields needed in this book. For details, see [79] or [49].

A locally compact non-discrete field is called a **local field**. Let F be a local field. If F is connected, we call it **archimedean**. Otherwise, we call it **nonarchimedean**. If a local field is archimedean, then it is isomorphic to $\mathbb{R}$ or $\mathbb{C}$. Any non-archimedean local field of characteristic 0 can be described as a finite algebraic extensions of $\mathbb{Q}_p$, for some p. Here, $\mathbb{Q}_p$ is the field of p-adic numbers defined below.

A.2.1   p-Adic Numbers

For an introduction to the p-adic numbers, see Schikhof [59] or Koblitz [41].

Definition A.6 Let F be a field.

(i) An **absolute value** on F is a map $\| \ \| : F \twoheadrightarrow \mathbb{R}_{\geq 0}$ such that, for all $x, y \in F$,

    (1) $\|x\| = 0$ if and only if $x = 0$,

    (2) $\|xy\| = \|x\| \cdot \|y\|$,

    (3) $\|x + y\| \leq \|x\| + \|y\|$ (the triangle inequality).

(ii) An absolute value $\| \ \|$ is called **nonarchimedean** if for all $x, y \in F$

    (3') $\|x + y\| \leq \max\{\|x\|, \|y\|\}$ (the strong triangle inequality).

Let x be a nonzero rational number. Then there exists a unique integer $\mathrm{ord}_p(x)$ such that $x = p^{\mathrm{ord}_p(x)} \dfrac{a}{b}$, where $p \nmid ab$. We define the p-**adic absolute value** of x by

$$|x|_p = p^{-\mathrm{ord}_p(x)}.$$

For $x = 0$, we define $|x|_p = 0$. Then $|\ |_p$ is a nonarchimedean absolute value on $\mathbb{Q}$. It induces the metric $d_p : \mathbb{Q} \times \mathbb{Q} \to \mathbb{R}_{\geq 0}$ given by $d_p(x, y) = |x - y|_p$ which satisfies the strong triangle inequality

$$d_p(x, y) \leq \max(d_p(x, z), d_p(z, y))$$

for all $x, y, z \in \mathbb{Q}$. Hence, $(\mathbb{Q}, d_p)$ is an ultrametric space. We denote by $\mathbb{Q}_p$ the completion of $\mathbb{Q}$ with respect to d_p. Then $\mathbb{Q}_p$ is also a field, called the **field of p-adic numbers**. The p-adic absolute value extends to $\mathbb{Q}_p$. Namely, if $x \in \mathbb{Q}_p$, then there exists a Cauchy sequence $\{x_n\}$ of rational numbers such that $x = \lim_{n \to \infty} x_n$. Then

$$|x|_p = \lim_{n \to \infty} |x_n|_p.$$

Define

$$\mathbb{Z}_p = \{x \in \mathbb{Q}_p \mid |x|_p \leq 1\}.$$

It is easy to show, using the strong triangle inequality, that $\mathbb{Z}_p$ is a ring, called the **ring of p-adic integers**. Note that $\mathbb{Z} \subset \mathbb{Z}_p$. The ring $\mathbb{Z}_p$ is a local ring, with the unique maximal ideal

$$\mathfrak{p} = (p) = p\mathbb{Z}_p = \{x \in \mathbb{Z}_p \mid |x|_p < 1\}.$$

The ring $\mathbb{Z}_p$ is compact and open. The ideals $\mathfrak{p}^n = p^n \mathbb{Z}_p$, $n \in \mathbb{N}$, form a neighborhood basis of zero consisting of compact open sets. For $a \in \mathbb{Q}_p$, the set $\{a + p^n \mathbb{Z}_p \mid n \in \mathbb{N}\}$ is a neighborhood basis of a consisting of compact open sets.

There is another construction of $\mathbb{Q}_p$, starting with the definition of the ring $\mathbb{Z}_p$. We consider the ring $\mathbb{Z}$ and the ideals (p^n) in $\mathbb{Z}$. We define $\mathbb{Z}_p$ as the projective limit

$$\mathbb{Z}_p = \varprojlim_n \mathbb{Z}/(p^n).$$

Then $\mathbb{Q}_p$ can be described algebraically as the field of fractions of $\mathbb{Z}_p$ or as

$$\mathbb{Q}_p = \mathbb{Z}_p \otimes_{\mathbb{Z}} \mathbb{Q} \quad \text{or as} \quad \mathbb{Q}_p = \bigcup_{n \geq 1} p^{-n}\mathbb{Z}_p.$$

As an ultrametric space, $\mathbb{Q}_p$ is totally disconnected. Let us mention that $\mathbb{Q}_p$ is not discrete. (A topological space X is discrete if each point in X is an open subset.) We will prove that the set $\{0\}$ is not open, which is equivalent to proving that $S = \mathbb{Q}_p \setminus \{0\}$ is not closed. Consider the Cauchy sequence $\{p^n\}$ in S. Then $\lim_{n \to \infty} p^n = 0 \notin S$, so S is not closed.

A.2.2   Finite Extensions of $\mathbb{Q}_p$

Let K be a finite extension of $\mathbb{Q}_p$ with $d = [K : \mathbb{Q}_p]$. For any $x \in K$, define

$$|x|_K = |N_{K/\mathbb{Q}_p}(x)|_p, \tag{A.1}$$

where the right-hand side is the p-adic absolute value on $\mathbb{Q}_p$. Then $|\ |_K$ is a nonarchimedean absolute value on K. Note that this absolute value does not extend $|\ |_p$. The absolute value which extends $|\ |_p$ is denoted again by $|\ |_p$ and it is computed as

$$|x|_p = |N_{K/\mathbb{Q}_p}(x)|_p^{1/d}, \quad x \in K. \tag{A.2}$$

The field K is totally disconnected and locally compact. Set

$$o_K = \{x \in K \mid |x|_K \leq 1\}.$$

It is easy to show that o_K is a ring, called the **ring of integers** of K. Its group of units is $o_K^\times = \{x \in K \mid |x|_K = 1\}$. The ring o_K is local, with the unique maximal ideal

$$\mathfrak{p}_K = \{x \in o_K \mid |x|_K < 1\}.$$

The ideal $\mathfrak{p}_K$ is principal. Every nonzero ideal of o_K is principal and is a power of $\mathfrak{p}_K$.

Any element of $\mathfrak{p}_K$ of maximal norm is a generator of $\mathfrak{p}_K$. Such an element is called a **uniformizer**. Fix a uniformizer ϖ_K. Then any nonzero $x \in K$ can be written as $x = \varpi_K^n u$, where $n \in \mathbb{Z}$ and $u \in o_K^\times$. The power n is denoted by $v_K(x)$ and is called the **valuation** of x. In addition, we define $v_K(0) = \infty$. We have

$$K = \bigcup_{n \in \mathbb{N}} \varpi_K^{-n} o_K.$$

If we define $d_K(x, y) = |x - y|_K$, for $x, y \in K$, we obtain an ultrametric $d_K : K \times K \to K$. Then (K, d_K) is an ultrametric space. The topology on K is algebraic in nature. It is easy to show that for any $0 < r < 1$, the ball of radius r centered at zero

$$B_r(0) = \{x \in K \mid |x|_K \le r\}$$

is an ideal in o_K, and hence it is of the form $\mathfrak{p}_K^n$ for some $n \in \mathbb{N}$. Then K has a neighborhood basis of zero consisting of the ideals $\mathfrak{p}_K^n$, $n \in \mathbb{N}$. Moreover, an arbitrary element $a \in K$ has a neighborhood basis of the form $a + \mathfrak{p}_K^n$, $n \in \mathbb{N}$.

Similarly, the topology on $o_K^\times$ is also algebraic in nature. For $n \in \mathbb{N}$, the set $1 + \mathfrak{p}_K^n$ is a subgroup of $o_K^\times$, and also an open neighborhood of 1 in $o_K^\times$.

Proposition A.7 *The canonical mapping*

$$o_K \longrightarrow \varprojlim_{n \in \mathbb{N}} o_K / \mathfrak{p}_K^n$$

is an isomorphism and a homeomorphism. The same is true for the mapping

$$o_K^\times \longrightarrow \varprojlim_{n \in \mathbb{N}} o_K^\times / (1 + \mathfrak{p}_K^n).$$

Proof This is Proposition 4.5 in Chapter II of Neukirch [49].   □

The **residue field** of K is defined as $\kappa = o_K / \mathfrak{p}_K$. The inclusion of $\mathbb{Z}_p$ into o_K induces a map of $\mathbb{F}_p = \mathbb{Z}_p / p\mathbb{Z}_p$ into κ and thus $\mathbb{F}_p$ is the prime field of κ. Let $f = [\kappa : \mathbb{F}_p]$. We call f the **residue degree** of K over $\mathbb{Q}_p$.

Denote by q_K the order of κ. Then $q_K = p^f$. We have $|\varpi_K|_K = q_K^{-1}$ and $|p|_K = p^{-d}$, where $d = [K : \mathbb{Q}_p]$. The absolute value of $x \in K^\times$ is

$$|x|_K = q_K^{-v_K(x)}.$$

Since any ideal in o_K is a power of $\mathfrak{p}_K$, the same is true for the principal ideal $(p) = p o_K \subset o_K$. It is of the form $\mathfrak{p}^e = (\varpi_K^e)$ for some $e \in \mathbb{N}$. The number e is called the **ramification degree** of K over $\mathbb{Q}_p$. We have $[K : \mathbb{Q}_p] = ef$ since

$$p^{-d} = |p|_K = |\varpi|_K^e = q_K^{-e} = p^{-ef}.$$

Lemma A.8 *Let K be a finite extension of $\mathbb{Q}_p$. Every strictly decreasing sequence in $|K|$ converges to zero.*

Proof This is clear since $|K| = |K^\times| \cup \{0\} = \{q_K^n \mid n \in \mathbb{Z}\} \cup \{0\}$. □

The above lemma is related to K being discretely valued. Namely, if F is a field with the absolute value $\| \|$, we say that F is **discretely valued** if the set $\|F^\times\|$ is discrete in $\mathbb{R}$.

A.2.3   Algebraic Closure $\overline{\mathbb{Q}}_p$

We denote by $\overline{\mathbb{Q}}_p$ the algebraic closure of $\mathbb{Q}_p$. It is used in Chap. 6 where we consider it as an abstract field.

It is known that two algebraically closed fields of the same characteristic and same uncountable cardinality are isomorphic [46, Proposition 2.2.5]. It follows that $\overline{\mathbb{Q}}_p$ and $\mathbb{C}$ are isomorphic as abstract fields.

The p-adic absolute value on $\mathbb{Q}_p$ extends to an absolute value on $\overline{\mathbb{Q}}_p$, denoted again by $|\,|_p$. The field $\overline{\mathbb{Q}}_p$ is not complete [41, Theorem 12]. We denote by $\mathbb{C}_p$ the completion of $\overline{\mathbb{Q}}_p$ with respect to $|\,|_p$. Then $\mathbb{C}_p$ is algebraically closed and complete [41, Theorem 13]. By the same arguments as above, $\mathbb{C}_p$ is also isomorphic to $\mathbb{C}$ as an abstract field.

A.3   Normed Vector Spaces

In this section, K is a finite extension of $\mathbb{Q}_p$ and $|\,| = |\,|_K$.

Definition A.9

(i) Let V be a K-vector space. A (nonarchimedean) **norm** on V is a map $\| \| : F \to \mathbb{R}_{\geq 0}$ such that

   (1) $\|v\| = 0$ if and only if $v = 0$,
   (2) $\|av\| = |a| \cdot \|v\|$, for $a \in K$, $v \in V$
   (3) $\|v + w\| \leq \max\{\|v\|, \|w\|\}$, for $v, w \in V$.

(ii) A **normed K-space** is a pair $(V, \| \|)$ consisting of a K-vector space V together with a norm $\| \|$ on V.

Let $(V, \| \|)$ be a normed K-space. The norm $\| \|$ induces a metric on V given by the formula $d(v, w) = \|v - w\|$. Then (V, d) is an ultrameteric space, as in Definition A.1.

Example A.10 Let $V = K^n$ be the n-dimensional K-vector space. Define

$$\|(a_1, \ldots, a_n)\| = \max_{1 \leq i \leq n} |a_i|.$$

This is a norm on V. The space $V = K^n$ is complete with respect to the corresponding metric, and hence it has the structure of a Banach space.

For more examples, see Sect. 3.1.3.

The **direct sum** $U \oplus V$ of two normed vector spaces U and V is the vector space $U \times V$ equipped with the norm

$$\|(u, v)\| = \max\{\|u\|, \|v\|\}. \tag{A.3}$$

If U and V are Banach spaces, then so is $U \oplus V$.

We conclude this section with the following simple and very useful result which follows from the isosceles triangle principle for ultrametric spaces.

Lemma A.11 *Let $(V, \| \ \|)$ be a normed K-space. If $v, w \in V$ satisfy $\|v\| \neq \|w\|$, then*

$$\|v + w\| = \max\{\|v\|, \|w\|\}.$$

Proof Follows directly from Proposition A.2, with $x = v$, $y = 0$, and $z = -w$. □

Appendix B
Affine and Projective Varieties

In this section, k is an algebraically closed field.

B.1 Affine Varieties

Let F be a field. The set $F^n = F \times \cdots \times F$ will be called **affine n-space over F** and denoted $\mathbf{A}^n$. A subset $\mathcal{V}$ of $\mathbf{A}^n$ is called an **affine variety** if $\mathcal{V}$ is the set of common zeros of a finite collection $S = \{f_\alpha\}$ of polynomials in $F[X] = F[x_1, \ldots, x_n]$. We write $\mathcal{V} = \mathcal{V}(S)$.

Example B.1 Some familiar objects in the real plane and the three-dimensional real space are affine varieties over $\mathbb{R}$.

(a) The circle $x^2 + y^2 = 1$ is the set of zeros of the polynomial $f(x, y) = x^2 + y^2 - 1$ in the two-dimensional affine space.
(b) The paraboloid $z = x^2 + y^2$ is the set of zeros of the polynomial $f(x, y, z) = x^2 + y^2 - z$ in the three-dimensional affine space.
(c) Let $f(x, y, z) = x^2 + y^2 - z^2$ and $g(x, y, z) = y + z - 1$. The affine variety $\mathcal{V}(f, g)$ is the intersection of the cone $z^2 = x^2 + y^2$ and the plane $z = 1 - y$, so it is a conic section (a parabola).

If I is an ideal of $F[X]$, we denote by $\mathcal{V}(I)$ the set of its common zeros in $\mathbf{A}^n$. By Hilbert's Basis Theorem, $F[X]$ is noetherian, so there exists a finite set of generators of I. This implies $\mathcal{V}(I)$ is an affine variety. Let $U \subset \mathbf{A}^n$. We denote by $I(U)$ the collection of all polynomials vanishing on U. Then $I(U)$ is an ideal. We have

$$U \subset \mathcal{V}(I(U)), \qquad I \subset I(\mathcal{V}(I)).$$

© The Author(s), under exclusive license to Springer Nature Switzerland AG 2022
D. Ban, *p-adic Banach Space Representations*, Lecture Notes
in Mathematics 2325, https://doi.org/10.1007/978-3-031-22684-7

Exercise B.2 Let $U, V \subset \mathbf{A}^n$ and $S, T \subset F[X]$. Prove:

$$U \subseteq V \Rightarrow \mathcal{I}(U) \supseteq \mathcal{I}(V), \quad \mathcal{I}(U \cup V) = \mathcal{I}(U) \cap \mathcal{I}(V),$$
$$S \subseteq T \Rightarrow \mathcal{V}(S) \supseteq \mathcal{V}(T), \quad \mathcal{V}(S \cup T) = \mathcal{V}(S) \cap \mathcal{V}(T). \tag{B.1}$$

From now on, $\mathbf{A}^n$ is the affine n-space over an algebraically closed field k. The **radical** of the ideal I is defined as

$$\text{rad } I = \{f \in k[X] \mid f^r \in I, \text{ for some } r \geq 0\}.$$

The ideal I is called a **radical ideal** if rad $I = I$.

Theorem B.3 (*Hilbert's Nullstellensatz*) *If I is an ideal in $k[x_1, \ldots, x_n]$, then*

$$\mathcal{I}(\mathcal{V}(I)) = \text{rad } I.$$

Moreover, the maps $\mathcal{V}$ and $\mathcal{I}$ in the correspondence

$$\{affine\ varieties\} \quad \overset{\mathcal{I}}{\underset{\mathcal{V}}{\rightleftarrows}} \quad \{radical\ ideals\}$$

are bijections that are inverses of each other.

Proof See Theorem 32 in Chapter 15 of [28]. □

Example B.4

(a) If $c = (c_1, \ldots, c_n) \in \mathbf{A}^n$, then $\mathcal{I}(c) = (x_1 - c_1, \ldots, x_n - c_n)$.
(b) Let $I \subset k[X]$ be a maximal ideal and $U = \mathcal{V}(I)$. From the Nullstellensatz, U is nonempty, so let $c = (c_1, \ldots, c_n) \in U$. Then (B.1) $I \subseteq \mathcal{I}(c) \neq k[X]$, so by maximality $I = \mathcal{I}(c)$ and $U = \mathcal{V}(I) = \mathcal{V}(\mathcal{I}(c)) = \{c\}$. There is a bijective correspondence between points in $\mathbf{A}^n$ and maximal ideals in $k[X]$.

B.1.1   Zariski Topology on Affine Space

Recall that a topology on a set X can be defined by a collection of closed subsets satisfying:

(T1)   The empty set and X are closed.
(T2)   The intersection of any collection of closed sets is also closed.
(T3)   The union of any finite number of closed sets is also closed.

We want to define the topology on $\mathbf{A}^n$ in which the closed sets are the affine varieties. We first observe that the empty set and $\mathbf{A}^n$ are affine varieties, so property (T1) holds. For (T2) and (T3), we apply the following exercise:

Exercise B.5

(a) If I and J are ideals in $k[X]$, prove that $\mathcal{V}(I) \cup \mathcal{V}(J) = \mathcal{V}(IJ)$.
(b) If $I_\alpha, \alpha \in \mathcal{A}$ is an arbitrary collection of ideals in $k[X]$, prove that

$$\bigcap_\alpha \mathcal{V}(I_\alpha) = \mathcal{V}(\sum_\alpha I_\alpha).$$

Then (a) implies (T3) and (b) implies (T2). It follows that affine sets define a topology on $\mathbf{A}^n$.

Definition B.6 The **Zariski topology** on affine n-space is the topology in which the closed sets are the affine varieties in $\mathbf{A}^n$.

Every point in $\mathbf{A}^n$ is a closed set. But the Hausdorff separation axiom fails. There are relatively few closed (or open) sets.

A nonempty affine variety V is called **irreducible** if it cannot be written as $V = V_1 \cup V_2$, where V_1 and V_2 are proper affine varieties.

Proposition B.7 *Let V be an affine variety.*

(i) V is irreducible if and only if $\mathcal{I}(V)$ is a prime ideal.
(ii) If V is not empty, it may be written uniquely in the form $V = V_1 \cup V_2 \cup \cdots \cup V_q$, where each V_i is irreducible and $V_i \not\subseteq V_j$ for $j \neq i$.

Proof Dummit and Foote [28, page 680].   □

B.1.2   Morphisms and Products of Affine Varieties

Suppose $V \subset \mathbf{A}^n$ and $W \subset \mathbf{A}^m$ are two affine varieties. A map $\varphi : V \to W$ is called a **morphism** of affine varieties if there are polynomials $\varphi_1, \ldots, \varphi_m \in k[x_1, \ldots, x_n]$ such that

$$\varphi((a_1, \ldots, a_n)) = (\varphi_1(a_1, \ldots, a_n), \ldots \varphi_m(a_1, \ldots, a_n)),$$

for all $(a_1, \ldots, a_n) \in V$. The map $\varphi : V \to W$ is an **isomorphism** of affine varieties if there is a morphism $\psi : W \to V$ with $\varphi \circ \psi = 1_W$ and $\psi \circ \varphi = 1_V$.

Note that in general $\varphi_1, \ldots, \varphi_m$ are not uniquely defined. Assume

$$(\varphi_1(a), \ldots \varphi_m(a)) = (\varphi_1'(a), \ldots \varphi_m'(a)),$$

for all $a = (a_1, \ldots, a_n) \in V$. Then, for $i = 1, \ldots, m$, $(\varphi_i - \varphi_i')(a) = 0$ for all $a \in V$, so $\varphi_i - \varphi_i' \in \mathcal{I}(V)$. We define

$$k[V] = k[X]/\mathcal{I}(V)$$

and call it the **affine algebra** of V (or the **algebra of polynomial functions** on V). The distinct polynomial functions on V are in one-to-one correspondence with $k[V]$. A morphism $\varphi : V \to W$ is a mapping of the form

$$\varphi(a) = (\psi_1(a), \ldots, \psi_m(a)),$$

where $\psi_i \in k[V]$. A morphism $\varphi : V \to W$ is continuous for the Zariski topologies involved. Indeed, if $U \subset V$ is the set of zeros of polynomial functions f_i on W, then $\varphi^{-1}(U)$ is the set of zeros of the polynomial functions $f_i \circ \varphi$ on V.

Let $V \subset \mathbf{A}^n$ and $U \subset \mathbf{A}^m$ be affine varieties. Then V is the common zeros of a finite set of polynomials $\{f_i \mid i = 1, \ldots, p\}$ in $k[x_1, \ldots, x_n]$, $V = \mathcal{V}(\{f_i\})$, and $U = \mathcal{V}(\{g_j\})$ is the common zeros of a finite set of polynomials $\{g_j \mid j = 1, \ldots, q\}$ in $k[y_1, \ldots, y_m]$. Then, the cartesian product $V \times U$ is an affine variety in $\mathbf{A}^{n+m}$. It is the set of common zeros of $\{f_i g_j \mid i = 1, \ldots, p, \ j = 1, \ldots, q\} \subset k[x_1, \ldots, x_n, y_1, \ldots, y_m]$,

$$V \times U = \mathcal{V}(\{f_i g_j\}) \subset \mathbf{A}^{n+m}.$$

We define the topology on $\mathbf{A}^n \times \mathbf{A}^m$, called the **Zariski product topology**, which identifies $\mathbf{A}^n \times \mathbf{A}^m$ with $\mathbf{A}^{n+m}$. This topology is different from the usual product topology.

Exercise B.8 Let $k = \mathbb{C}$.

(a) Prove that the Zariski-closed sets in $\mathbf{A}^1$ are $\emptyset$, $\mathbb{C}$, and all finite sets.
(b) Prove that the Zariski product topology on $\mathbf{A}^1 \times \mathbf{A}^1$ (which identifies it with $\mathbf{A}^2$) is strictly finer than the product topology on $\mathbf{A}^1 \times \mathbf{A}^1$.

B.2   Projective Varieties

Projective n-space $\mathbf{P}^n$ is the set of equivalence classes of $k^{n+1} - \{(0, \ldots, 0)\}$ relative to the equivalence relation

$$(c_0, c_1, \ldots, c_n) \sim (d_0, d_1, \ldots, d_n)$$

if and only if

$$(d_0, d_1, \ldots, d_n) = \alpha(c_0, c_1, \ldots, c_n) = (\alpha c_0, \alpha c_1, \ldots, \alpha c_n),$$

for some $\alpha \in k^\times$. Each point in $\mathbf{P}^n$ can be described by **homogeneous coordinates** $c_0, c_1, \ldots, c_n$ which are not unique but may be multiplied by any nonzero scalar.

A polynomial $f \in k[X] = K[x_0, \ldots, x_n]$ is **homogeneous** of degree d if it is a linear combination of monomials of degree d. This is equivalent to the condition

$$f(\alpha x_0, \ldots, \alpha x_n) = \alpha^d f(x_0, \ldots, x_n),$$

$\alpha \in k^\times$. If $f \in k[X]$ is a homogeneous polynomial and $f(c_0, c_1, \ldots, c_n) = 0$, then

$$f(\alpha c_0, \alpha c_1, \ldots, \alpha c_n) = \alpha^d f(c_0, \ldots, c_n) = 0.$$

This justifies the following definition. A subset $\mathcal{V}$ of $\mathbf{P}^n$ is called a **projective variety** if $\mathcal{V}$ is the set of common zeros of a finite collection $S = \{f_\alpha\}$ of homogeneous polynomials in $K[X]$. We write $\mathcal{V} = \mathcal{V}(S)$. Let $U \subset \mathbf{P}^n$. We denote by $\mathcal{I}(U)$ the collection of all polynomials vanishing on U. Then $\mathcal{I}(U)$ is an ideal.

An arbitrary polynomial $f \in k[X]$ can be written in the form

$$f = \sum f^{(d)},$$

where $f^{(d)} \in k[X]$ is a homogeneous polynomial of degree d. We call $f^{(d)}$ the **homogeneous part** of f of degree d. An ideal I of $k[X]$ is called **homogeneous** if whenever $f \in I$ then each homogeneous part $f^{(d)}$ also lies in I.

Exercise B.9 Let U be a subset of projective space $\mathbf{P}^n$. Prove that the set $\mathcal{I}(U)$ of all polynomials vanishing on U is a homogeneous ideal.

Now we define a topology on $\mathbf{P}^n$ by taking a closed set to be a projective variety (the common zeros of a collection of homogeneous polynomials, or equivalently of the ideal they generate).

Let I_0 denote the ideal generated by $x_0, \ldots, x_n$. Then I_0 has no common zeros in $\mathbf{P}^n$.

Proposition B.10 *The operators* $\mathcal{V}, \mathcal{I}$ *set a one-to-one inclusion-reversing correspondence between the closed subsets of* $\mathbf{P}^n$ *and the homogeneous radical ideals of* $k[X]$ *other than* I_0.

Proof This is Proposition 1.6 in Humphreys [38].   □

References

1. N. Abe, G. Henniart, F. Herzig, M.-F. Vignéras, A classification of irreducible admissible mod p representations of p-adic reductive groups. J. Am. Math. Soc. **30**(2), 495–559 (2017)
2. K. Ardakov, K.A. Brown, Ring-theoretic properties of Iwasawa algebras: a survey. Documenta Mathematica, Extra Vol., John H. Coates' Sixtieth Birthday, 7–33 (2006)
3. D. Ban, J. Hundley, On reducibility of p-adic principal series representations of p-adic groups. Represent. Theory **20**, 249–262 (2016)
4. D. Ban, J. Hundley, Intertwining maps between p-adic principal series of p-adic groups. Represent. Theory **25**, 975–993 (2021)
5. L. Berger, Représentations modulaires de $GL_2(\mathbf{Q}_p)$ et représentations galoisiennes de dimension 2. Astérisque **330**, 263–279 (2010)
6. L. Berger, La correspondance de Langlands locale p-adique pour $GL_2(\mathbf{Q}_p)$. Astérisque, no. 339, Exp. No. 1017, viii, 157–180, Séminaire Bourbaki. Vol. 2009/2010. Exposés 1012–1026 (2011)
7. L. Berger, C. Breuil, Sur quelques représentations potentiellement cristallines de $GL_2(\mathbf{Q}_p)$. Astérisque **330**, 155–211 (2010)
8. A. Borel, *Linear Algebraic Groups*. Graduate Texts in Mathematics, vol. 126, 2nd edn. (Springer, New York, 1991)
9. A. Borel, J. Tits, Groupes réductifs. Inst. Hautes Études Sci. Publ. Math. **27**, 55–150 (1965)
10. N. Bourbaki, *General Topology. Chapters 5–10*. Elements of Mathematics (Berlin) (Springer, Berlin, 1998). Translated from the French, Reprint of the 1989 English translation
11. C. Breuil, Sur quelques représentations modulaires et p-adiques de $GL_2(\mathbf{Q}_p)$. I. Compos. Math. **138**(2), 165–188 (2003)
12. C. Breuil, P. Schneider, First steps towards p-adic Langlands functoriality. J. Reine Angew. Math. **610**, 149–180 (2007)
13. F. Bruhat, J. Tits, Groupes réductifs sur un corps local. Inst. Hautes Études Sci. Publ. Math. **41**, 5–251 (1972)
14. C.J. Bushnell, G. Henniart, *The Local Langlands Conjecture for* GL(2). Grundlehren der Mathematischen Wissenschaften [Fundamental Principles of Mathematical Sciences], vol. 335 (Springer, Berlin, 2006)
15. A. Caraiani, M. Emerton, T. Gee, D. Geraghty, V. Paškūnas, S.W. Shin, Patching and the p-adic local Langlands correspondence. Camb. J. Math. **4**(2), 197–287 (2016)
16. P. Cartier, Representations of p-adic groups: a survey, in *Automorphic Forms, Representations and L-Functions (Proceedings of Symposia in Pure Mathematics, Oregon State University, Corvallis, Oregon, 1977), Part 1*. Proceedings of Symposium Pure Mathematics, vol. 33 (American Mathematical Society, Providence, 1979), pp. 111–155

© The Author(s), under exclusive license to Springer Nature Switzerland AG 2022

D. Ban, *p-adic Banach Space Representations*, Lecture Notes
in Mathematics 2325, https://doi.org/10.1007/978-3-031-22684-7

17. W. Casselman, Introduction to the theory of admissible representations of p-adic reductive groups. https://www.math.ubc.ca/~cass/research/pdf/p-adic-book.pdf

18. W. Casselman, The unramified principal series of p-adic groups I. The spherical function. Compos. Math. **40**(3), 387–406 (1980)

19. C. Chevalley, Certains schémas de groupes semi-simples, in *Séminaire Bourbaki*, vol. 6 (Society of Mathematics, France, 1995), pp. 219–234

20. P. Colmez, Représentations de $GL_2(\mathbf{Q}_p)$ et (ϕ, Γ)-modules. Astérisque **330**, 281–509 (2010)

21. P. Colmez, La série principale unitaire de $GL_2(\mathbf{Q}_p)$: vecteurs localement analytiques, in *Automorphic Forms and Galois Representations. Vol. 1*. London Mathematical Society Lecture Note Series, vol. 414 (Cambridge University Press, Cambridge, 2014), pp. 286–358

22. P. Colmez, G. Dospinescu, V. Paškūnas, The p-adic local Langlands correspondence for $GL_2(\mathbb{Q}_p)$. Camb. J. Math. **2**(1), 1–47 (2014)

23. B. Conrad, Reductive group schemes, in *Autour des schémas en groupes. Vol. I*. Panoramas Synthèses, vols. 42–43 (Society of Mathematics, France, 2014), pp. 93–444

24. C.W. Curtis, I. Reiner, *Representation Theory of Finite Groups and Associative Algebras* (AMS Chelsea Publishing, Providence, 2006). Reprint of the 1962 original

25. M. Demazure, P. Gabriel, *Groupes algébriques. Tome I: Géométrie algébrique, généralités, groupes commutatifs* (Masson & Cie, Éditeur; North-Holland, Amsterdam, 1970). Avec un appendice ıt Corps de classes local par Michiel Hazewinkel

26. M. Demazure, A. Grothendieck (Eds.), Schémas en groupes. III: Structure des schémas en groupes réductifs, in *Séminaire de Géométrie Algébrique du Bois Marie 1962/64 (SGA 3)*. Lecture Notes in Mathematics, vol. 153 (Springer, Berlin, 1970)

27. B. Diarra, Sur quelques représentations p-adiques de $\mathbf{Z}_p$. Nederl. Akad. Wetensch. Indag. Math. **41**(4), 481–493 (1979)

28. D.S. Dummit, R.M. Foote, *Abstract Algebra*, 3rd edn. (Wiley, Hoboken, 2004)

29. M. Emerton, p-adic L-functions and unitary completions of representations of p-adic reductive groups. Duke Math. J. **130**(2), 353–392 (2005)

30. M. Emerton, A local-global compatibility conjecture in the p-adic Langlands programme for $GL_{2/\mathbb{Q}}$. Pure Appl. Math. Q. **2**, 279–393 (2006)

31. M. Emerton, Ordinary parts of admissible representations of p-adic reductive groups I. Definition and first properties. Astérisque **331**, 355–402 (2010)

32. M. Emerton, Locally analytic vectors in representations of locally p-adic analytic groups. Mem. Am. Math. Soc. **248**(1175), iv+158 (2017)

33. H. Frommer, The locally analytic principal series of split reductive groups. Preprintreihe SFB 478, Münster, Heft 265 (2003)

34. I.M. Gel'fand, M.I. Graev, I.I. Pyatetskii-Shapiro, *Representation Theory and Automorphic Functions*. Generalized Functions, vol. 6 (Academic Press, Boston, 1990). Translated from the Russian by K. A. Hirsch, Reprint of the 1969 edition

35. P. Gille, P. Polo (Eds.), *Schémas en groupes (SGA 3). Tome I. Propriétés générales des schémas en groupes*. Documents Mathématiques (Paris) [Mathematical Documents (Paris)], vol. 7 (Société Mathématique de France, France, 2011). Séminaire de Géométrie Algébrique du Bois Marie 1962–64. [Algebraic Geometry Seminar of Bois Marie 1962–64]. A seminar directed by M. Demazure and A. Grothendieck with the collaboration of M. Artin, J.-E. Bertin, P. Gabriel, M. Raynaud and J-P. Serre, Revised and annotated edition of the 1970 French original

36. Harish-Chandra, *Admissible Invariant Distributions on Reductive p-Adic Groups*. University Lecture Series, vol. 16 (American Mathematical Society, Providence, 1999). With a preface and notes by Stephen DeBacker and Paul J. Sally, Jr.

37. G. Henniart, M.-F. Vignéras, Representations of a p-adic group in characteristic p, in *Representations of Reductive Groups*. Proceedings of Symposia in Pure Mathematics, vol. 101 (American Mathematical Society, Providence, 2019), pp. 171–210

38. J.E. Humphreys, *Linear Algebraic Groups* (Springer, New York, 1975). Graduate Texts in Mathematics, No. 21

39. J.E. Humphreys, *Introduction to Lie Algebras and Representation Theory*. Graduate Texts in Mathematics, vol. 9 (Springer, New York, 1978). Second printing, revised
40. J.C. Jantzen, *Representations of Algebraic Groups*. Mathematical Surveys and Monographs, vol. 107, 2nd edn. (American Mathematical Society, Providence, 2003)
41. N. Koblitz, *p-Adic Numbers, p-adic Analysis, and Zeta-Functions*. Graduate Texts in Mathematics, vol. 58, 2nd edn. (Springer, New York, 1984)
42. T.Y. Lam, *A First Course in Noncommutative Rings*. Graduate Texts in Mathematics, vol. 131, 2nd edn. (Springer, New York, 2001)
43. S. Lang, *Algebra*. Graduate Texts in Mathematics, vol. 211, 3rd edn. (Springer, New York, 2002)
44. M. Lazard, Groupes analytiques *p*-adiques. Inst. Hautes Études Sci. Publ. Math. **26**, 389–603 (1965)
45. S. Mac Lane, *Categories for the Working Mathematician*. Graduate Texts in Mathematics, vol. 5, 2nd edn. (Springer, New York, 1998)
46. D. Marker, *Model Theory*. Graduate Texts in Mathematics, vol. 217 (Springer, New York, 2002). An introduction
47. A.F. Monna, T.A. Springer, Intégration non-archimédienne I. Nederl. Akad. Wetensch. Proc. Ser. A 66=Indag. Math. **25**, 634–642 (1963)
48. J.R. Munkres, *Topology* (Prentice Hall, Upper Saddle River, 2000). Second edition of [MR0464128]
49. J. Neukirch, *Algebraic Number Theory*. Grundlehren der Mathematischen Wissenschaften [Fundamental Principles of Mathematical Sciences], vol. 322 (Springer, Berlin, 1999). Translated from the 1992 German original and with a note by Norbert Schappacher, With a foreword by G. Harder
50. S. Orlik, M. Strauch, On the irreducibility of locally analytic principal series representations. Represent. Theory **14**, 713–746 (2010)
51. S. Orlik, M. Strauch, On Jordan-Hölder series of some locally analytic representations. J. Am. Math. Soc. **28**(1), 99–157 (2015)
52. V. Paškūnas, The image of Colmez's Montreal functor. Publ. Math. Inst. Hautes Études Sci. **118**, 1–191 (2013)
53. D. Renard, *Représentations des groupes réductifs p-adiques*. Cours Spécialisés [Specialized Courses], vol. 17 (Société Mathématique de France, France, 2010)
54. L. Ribes, P. Zalesskii, *Profinite Groups*. Ergebnisse der Mathematik und ihrer Grenzgebiete. 3. Folge. A Series of Modern Surveys in Mathematics [Results in Mathematics and Related Areas. 3rd Series. A Series of Modern Surveys in Mathematics], vol. 40, 2nd edn. (Springer, Berlin, 2010)
55. E. Riehl, *Category Theory in Context* (Dover Publication, Mineola, 2016)
56. J.J. Rotman, *An Introduction to Homological Algebra*. Pure and Applied Mathematics (Academic Press, New York, 1979)
57. W.H. Schikhof, The *p*-adic bounded weak topologies, in *Mathematical Contributions in Memory of Professor Victor Manuel Onieva Aleixandre (Spanish)* (University of Cantabria, Santander, 1991), pp. 293–300
58. W.H. Schikhof, A perfect duality between *p*-adic Banach spaces and compactoids. Indag. Math. **6**(3), 325–339 (1995)
59. W.H. Schikhof, *Ultrametric Calculus*. Cambridge Studies in Advanced Mathematics, vol. 4 (Cambridge University Press, Cambridge, 2006). An introduction to *p*-adic analysis, Reprint of the 1984 original [MR0791759]
60. P. Schneider, *Nonarchimedean Functional Analysis*. Springer Monographs in Mathematics (Springer, Berlin, 2002)
61. P. Schneider, Continuous representation theory of *p*-adic Lie groups, in *International Congress of Mathematicians*, vol. II (European Mathematical Society, Zürich, 2006), pp. 1261–1282
62. P. Schneider, *p*-adic Banach space representations of *p*-adic groups. Lectures at Jerusalem, March 30—April 6 (2009)

63. P. Schneider, *p-Adic Lie Groups*. Grundlehren der Mathematischen Wissenschaften [Fundamental Principles of Mathematical Sciences], vol. 344 (Springer, Heidelberg, 2011)

64. P. Schneider, J. Teitelbaum, Banach space representations and Iwasawa theory. Isr. J. Math. **127**, 359–380 (2002)

65. P. Schneider, J. Teitelbaum, Locally analytic distributions and *p*-adic representation theory, with applications to GL_2. J. Am. Math. Soc. **15**(2), 443–468 (2002)

66. P. Schneider, J. Teitelbaum, Algebras of *p*-adic distributions and admissible representations. Invent. Math. **153**(1), 145–196 (2003)

67. P. Schneider, J. Teitelbaum, Continuous and locally analytic representation theory. Lectures at Hangzhou (2004)

68. P. Schneider, J. Teitelbaum, D. Prasad, $U(\mathfrak{g})$-finite locally analytic representations. Represent. Theory **5**, 111–128 (2001). With an appendix by Dipendra Prasad

69. F. Shahidi, *Eisenstein Series and Automorphic L-Functions*. American Mathematical Society Colloquium Publications, vol. 58 (American Mathematical Society, Providence, 2010)

70. A.J. Silberger, *Introduction to Harmonic Analysis on Reductive p-Adic Groups*. Mathematical Notes, vol. 23 (Princeton University Press, Princeton; University of Tokyo Press, Tokyo, 1979). Based on lectures by Harish-Chandra at the Institute for Advanced Study, 1971–1973

71. T.A. Springer, Reductive groups, in *Automorphic Forms, Representations and L-Functions (Proceedings of Symposia in Pure Mathematics, Oregon State University, Corvallis, 1977), Part 1*. Proceedings of Symposia in Pure Mathematics, vol. 33 (American Mathematical Society, Providence, 1979), pp. 3–27

72. T.A. Springer, *Linear Algebraic Groups*. Progress in Mathematics, vol. 9, 2nd edn. (Birkhäuser, Boston, 1998)

73. R. Steinberg, Torsion in reductive groups. Adv. Math. **15**, 63–92 (1975)

74. M. Tadić, Notes on representations of non-Archimedean SL(*n*). Pac. J. Math. **152**(2), 375–396 (1992)

75. R. Taylor, Galois representations. Ann. Fac. Sci. Toulouse Math. **13**(1), 73–119 (2004)

76. A.V. Trusov, Representations of the groups GL(2, $\mathbf{Z}_p$) and GL(2, $\mathbf{Q}_p$) in spaces over non-Archimedian fields. Vestnik Moskov. Univ. Ser. I Mat. Mekh. **1**, 55–59, 108 (1981)

77. A.C.M. van Rooij, *Non-Archimedean Functional Analysis*. Monographs and Textbooks in Pure and Applied Mathematics, vol. 51 (Marcel Dekker, New York, 1978)

78. M.-F. Vignéras, *Représentations l-modulaires d'un groupe réductif p-adique avec l $\neq$ p*. Progress in Mathematics, vol. 137 (Birkhäuser, Boston, 1996)

79. A. Weil, *Basic Number Theory*. Classics in Mathematics (Springer, Berlin, 1995). Reprint of the second (1973) edition

80. A.V. Zelevinsky, Induced representations of reductive p-adic groups II. On irreducible representations of GL(*n*). Ann. Sci. École Norm. Sup. **13**(2), 165–210 (1980)

Index

© The Author(s), under exclusive license to Springer Nature Switzerland AG 2022
D. Ban, *p-adic Banach Space Representations*, Lecture Notes
in Mathematics 2325, https://doi.org/10.1007/978-3-031-22684-7

LECTURE NOTES IN MATHEMATICS   🐴 Springer

Editors in Chief: J.-M. Morel, B. Teissier;

Editorial Policy

1. Lecture Notes aim to report new developments in all areas of mathematics and their applications – quickly, informally and at a high level. Mathematical texts analysing new developments in modelling and numerical simulation are welcome.

   Manuscripts should be reasonably self-contained and rounded off. Thus they may, and often will, present not only results of the author but also related work by other people. They may be based on specialised lecture courses. Furthermore, the manuscripts should provide sufficient motivation, examples and applications. This clearly distinguishes Lecture Notes from journal articles or technical reports which normally are very concise. Articles intended for a journal but too long to be accepted by most journals, usually do not have this "lecture notes" character. For similar reasons it is unusual for doctoral theses to be accepted for the Lecture Notes series, though habilitation theses may be appropriate.

2. Besides monographs, multi-author manuscripts resulting from SUMMER SCHOOLS or similar INTENSIVE COURSES are welcome, provided their objective was held to present an active mathematical topic to an audience at the beginning or intermediate graduate level (a list of participants should be provided).

   The resulting manuscript should not be just a collection of course notes, but should require advance planning and coordination among the main lecturers. The subject matter should dictate the structure of the book. This structure should be motivated and explained in a scientific introduction, and the notation, references, index and formulation of results should be, if possible, unified by the editors. Each contribution should have an abstract and an introduction referring to the other contributions. In other words, more preparatory work must go into a multi-authored volume than simply assembling a disparate collection of papers, communicated at the event.

3. Manuscripts should be submitted either online at www.editorialmanager.com/lnm to Springer's mathematics editorial in Heidelberg, or electronically to one of the series editors. Authors should be aware that incomplete or insufficiently close-to-final manuscripts almost always result in longer refereeing times and nevertheless unclear referees' recommendations, making further refereeing of a final draft necessary. The strict minimum amount of material that will be considered should include a detailed outline describing the planned contents of each chapter, a bibliography and several sample chapters. Parallel submission of a manuscript to another publisher while under consideration for LNM is not acceptable and can lead to rejection.

4. In general, **monographs** will be sent out to at least 2 external referees for evaluation.

   A final decision to publish can be made only on the basis of the complete manuscript, however a refereeing process leading to a preliminary decision can be based on a pre-final or incomplete manuscript.

   Volume Editors of **multi-author works** are expected to arrange for the refereeing, to the usual scientific standards, of the individual contributions. If the resulting reports can be forwarded to the LNM Editorial Board, this is very helpful. If no reports are forwarded or if other questions remain unclear in respect of homogeneity etc, the series editors may wish to consult external referees for an overall evaluation of the volume.

5. Manuscripts should in general be submitted in English. Final manuscripts should contain at least 100 pages of mathematical text and should always include

   – a table of contents;
   – an informative introduction, with adequate motivation and perhaps some historical remarks: it should be accessible to a reader not intimately familiar with the topic treated;
   – a subject index: as a rule this is genuinely helpful for the reader.
   – For evaluation purposes, manuscripts should be submitted as pdf files.

6. Careful preparation of the manuscripts will help keep production time short besides ensuring satisfactory appearance of the finished book in print and online. After acceptance of the manuscript authors will be asked to prepare the final LaTeX source files (see LaTeX templates online: https://www.springer.com/gb/authors-editors/book-authors-editors/manuscriptpreparation/5636) plus the corresponding pdf- or zipped ps-file. The LaTeX source files are essential for producing the full-text online version of the book, see http://link.springer.com/bookseries/304 for the existing online volumes of LNM). The technical production of a Lecture Notes volume takes approximately 12 weeks. Additional instructions, if necessary, are available on request from lnm@springer.com.

7. Authors receive a total of 30 free copies of their volume and free access to their book on SpringerLink, but no royalties. They are entitled to a discount of 33.3 % on the price of Springer books purchased for their personal use, if ordering directly from Springer.

8. Commitment to publish is made by a *Publishing Agreement*; contributing authors of multiauthor books are requested to sign a *Consent to Publish form*. Springer-Verlag registers the copyright for each volume. Authors are free to reuse material contained in their LNM volumes in later publications: a brief written (or e-mail) request for formal permission is sufficient.

Addresses:
Professor Jean-Michel Morel, CMLA, École Normale Supérieure de Cachan, France
E-mail: moreljeanmichel@gmail.com

Professor Bernard Teissier, Equipe Géométrie et Dynamique,
Institut de Mathématiques de Jussieu – Paris Rive Gauche, Paris, France
E-mail: bernard.teissier@imj-prg.fr

Springer: Ute McCrory, Mathematics, Heidelberg, Germany,
E-mail: lnm@springer.com

Printed in the United States
by Baker & Taylor Publisher Services